Systems

Volume 4

Measuring Organisational Efficiency

Volume 1
A Journey Through the Systems Landscape
Harold "Bud" Lawson

Volume 2
A Discipline of Mathematical Systems Modelling
Matthew Collinson, Brian Monahan, and David Pym

Volume 3
Beyond Alignment. Applying Systems Thinking in Architecting Enterprises
John Gøtze and Anders Jensen-Waud, eds.

Volume 4
Measuring Organisational Efficiency
Francisco Parra-Luna and Eva Kasparova, in collaboration with Micael Frenck

Measuring Organisational Efficiency

Francisco Parra-Luna
and
Eva Kasparova

In collaboration with Micael Frenck

ISBN 978-1-84890-126-1

College Publications
Scientific Director: Dov Gabbay
Managing Director: Jane Spurr

Editorial Assistant: Asimina Koukou
Cover design adapted from an image by David Horvath, Chaoscope 0.3.1

http://www.collegepublications.co.uk

Printed by Lightning Source, UK

The Systems series publishes books related to Systems Science, Systems Thinking, Systems Engineering and Software Engineering.

Systems Science having its contemporary roots in the first half of the 20th century is today made up of a diversity of approaches hat have entered different fields of investigation. Systems Science explores how common features manifest in natural and social systems of varying complexity in order to provide scientific foundations for describing, understanding and designing systems.

Systems Thinking has grown during the latter part of the 20th century into highly useful discipline independent methods, languages and practices. Systems Thinking focuses upon applying concepts, principles, and paradigms in the analysis of the holistic structural and behavioral properties of complex systems. In particular the patterns of relationships that arise in the interactions of multiple systems.

Systems and Software Engineering. Systems Engineering has gained momentum during the latter part of the 20th century and has led to engineering related practices and standards that can be used in the life cycle management of complex systems. Software Engineering has continued to grow in importance as the software content of most complex systems has steadily increased and in many cases have become the dominant elements. Both Systems and Software Engineering focus upon transforming the need for a system into products and services hat meet the need in an effective, reliable and cost effective manner. While there are similarities between Systems and Software Engineering, the unique properties of software often requires special expertise and approaches to life cycle management.

Systems Science, Systems Thinking, as well as Systems and Software Engineering can, and need to, be considered complementary in establishing the capability to individually and collectively "think" and "act" in terms of systems in order to face the complex challenges of modern systems.

This series is a cooperative enterprise between College Publications, the School of Systems and Enterprises at Stevens Institute of Technology and the Bertalanffy Centre for the Study of Systems Science (BCSSS).

Contents

1

Introduction

By Francisco Parra Luna and Eva Kasparova

The first difficulty encountered when addressing the concept of "organisational efficiency" is its impressive polysemy, for it is related to, and often mistaken for, a whole series of terms the semantic boundaries of which have not been clearly defined. The result is that these concepts are regarded as synonymous and used indistinctly. Some may be briefly defined as they appear in specialised literature. One such example is the term "performance" (which is defined as recurrent activities designed to establish organisational goals and make adjustments to achieve those goals more effectively and efficiently); efficacity (ability to produce the desired results); efficacy (the individual's perceived expectations of obtaining desired outcomes through personal effort); effectiveness (the ability of an organisation to fulfil its mission through sound management, strong governance and a concentrated effort to achieve certain results); development (change in people and organisations for positive growth); productivity (output per unit of input); quality improvement (drive for excellence, often referring to education systems); excellence (the state or quality of supreme goodness for many kinds of organisations); success (the achievement of an organisational goal); profitability (the ability of an organisation to generate earnings for the benefit of its owners); evolution (gradual transformation of structures); and growth (an increase in some quantity over time). All these terms are often preceded by the word "organisational".

Moreover, the term "efficiency" is used to describe many different things in many different contexts or for many different applications. Hence, it may refer to "textual efficiency" (pithiness of a message); "material efficiency" (comparison of material requirements between construction projects or physical processes);

"business efficiency" (expenses as a percentage of revenues); "mechanical efficiency" (the effectiveness of a simple machine, or actual output over designed output); "energy efficiency" (useful work per unit of energy); "economic efficiency" (a general term to assess the amount of waste or other undesirable features); "allocating efficiency" (optimal distribution of goods); "wage efficiency" (paying workers more than the market rate for increased productivity); and so on. "Inefficiency", analogously, is used to mean the opposite in each context. A further series of related concepts will be discussed below in Chapter 8.

Finally, "organisational efficiency" has been defined (Ghemawat & Ricart, 1993) in two different ways: "in terms of the refinement of existing products, processes or capabilities (static efficiency) and as the development of new ones (dynamic efficiency)".

In other words, both definitions refer to "capacity" or "potential", but not to actual outcome or achievement. However, according to the definition that we have adopted in this book, an organisation is regarded as efficient if it is able to demonstrate its real and empirical capacity and not merely its theoretical potential.

First, then, the object of assessment should be final results rather than potential capabilities. What would be an exact definition of efficiency from this perspective? In principle, as noted, an organisation is efficient if it meets at least two pre-requisites: a) if it achieves its goals; and b) if it does so better and at a lower cost than the average organisations of its kind, whether public or private.

Secondly, the present volume does not address the processes that usually explain organisational success, but rather the description of performance, the fundamental problems of measuring such processes, the analysis of their conceptual dimensions, and the determination of the sum of their results. Our aim is therefore more modest: we want to offer an operational and quantitative definition of "organisational efficiency", as inclusive as possible to encompass its key dimensions. Consequently, this book will not focus on the explanatory aspects of "organisational efficiency"-related concepts such as "cultural change", "ISO9000 as an international standard of quality", "knowledge management", "learning organisation", "management by objectives (MBO)", "outcome-based evaluation", "programme evaluation", "strategic planning", "total quality management (TQM)" or others normally used in "team-building", "career development", "training", "e-learning", "coaching", "talent management" or "leadership development".

Thirdly, our definition of "organisational efficiency" will focus on and stress the value the system has generated. Efficiency will not be assessed in economic terms only, but also in social, ethical and environmental terms. Therefore, the operational definition will be based on what has been called the Referential Pattern of Values (RPV), which attaches as much weight and importance to the health status of employees as to economic performance; to the number of units and services produced per unit of time as to environmental quality; to the participation and personal development of workers as to their physical and occupational safety. This axiological approach is justified because organisations are essentially a group of persons who have, and are driven by, certain needs. Therefore, an organisation's sole obligation is to satisfy those needs, bearing in mind that the persons involved are not only employees, executives, citizens or shareholders, but also creditors, suppliers, customers and many other stakeholders who, at one time or another, depend on or are related in some way to the organisation, be it "profit-making" or "non-profit–making".

Organisational efficiency will, therefore, be considered as a global concept in three dimensions: global because it takes into account all the fundamental values pursued; global because it encompasses all the organisation's stakeholders; and global because it attempts to integrate other theoretical concepts related to the term "efficiency", primarily "efficacy" and "effectiveness". It is absolutely unrelated to ideologically charged notions such as Nazi Germany's "war efficiency" or "efficient mass annihilation", to name a few. Nor will the term "efficiency" be used here in the sense of "economic profit". The notion of organisational efficiency as used here merely considers economic issues as one dimension among several others.

Nonetheless, since our ultimate purpose is to measure efficiency in one way or another, all the authors contributing to this book have applied Lazarsfeld's (1965) well-known rule: a) passage from the image of a concept to the definition of its theoretical dimensions; b) subdivision of each dimension into a series of empirical indicators, which are either objective (statistical data based on facts), or subjective (opinions expressed in the organisation's social spheres – employees, clients, and so on, also expressed in quantitative terms); and c) the sum or integration of these indicators (weighted or otherwise by their relative importance) as the expression of the final result or quantification of the initial conceptual image.

This passage from "image to number" is one of the processes that enable social scientists to speak safely, albeit humbly, about what has been validated by data by means of methods that can be described. This empirical content of the concept may not match everyone's semantic image, but social scientists contend that they can base their assertions on existing validated data only. That is their ethical position.

Nevertheless, the concept of "organisational efficiency" addressed in this book has been defined from different perspectives and applied to different organisational dimensions and circumstances. The authors agree on a basic definition of organisational efficiency. That is: the capacity of an organisation to maximize certain desired outputs with a minimum input. In a dynamic context, efficiency should always seek the highest output/input ratio, assuming output growth or stabilisation. The purpose is to provide alternative theoretical angles that may help understand the concept from a practical, but above all, a humanistic point of view. Moreover, the term should globally cover a series of theoretical dimensions whose operational definition is highly demanding in terms of competence-related content. What this means is that companies labelled as "efficient" under this approach (i.e., having an overall index greater than one, as discussed below) will be able to assert that they are efficient now and will continue to be so in the future,,and that such efficiency is measured by comparison with their competitors.

Ultimately, given the predominant role of private enterprise in modern society, the concept will hinge on business efficiency, the backbone of any national economy, but the new concept stands at quite a distance from the narrow perspective of economic efficiency.

Given these assumptions, the questions that arise are: can "organisational efficiency" be measured? And what sort of information would be needed to achieve valid measurement?

Providing an appropriate or scientifically valid answer to these two questions involves solving certain basic theoretical problems. The first consists in practically denying a more or less accepted hypothesis according to which the concept cannot be measured. This hypothesis can be found throughout scientific literature on organisations, where it is claimed that "efficiency in organisations cannot be measured or calibrated for want of a general comparative model." This premise has been supported by most scholars who addressed the subject (Edwards et al., 1986).

Nonetheless, many authors have attempted to measure organisational efficiency empirically. Miles (1980), for instance, used 29 measurements; Campbell (1977) 30 criteria; Mahoney (1977) 114 variables; and Seashore & Yutchman (1967) 76 different indicators. Some authors (Dalton and Kesner, 1985) even claim that the number of possible measurements is nearly infinite, while all stress the difficulty involved in standardising measures for comparison. Generally speaking, positions range from proposals of a moratorium in the analysis of organisational efficiency until better inter-subjective conditions are in place (such as Goodman, Atkin and Schoormann, 1983), to those who advocate definitively abandoning the idea in view of the utter impossibility of ever reaching an agreement (such as Hannan & Freemann, 1977).

Other authors (such as Morgan, 1980) believe that such an agreement is not impossible or maintain that the decisive importance of the concept is such that it should not be abandoned if the aim pursued is to understand and improve business organisations (Peters & Waterman, 1982; Handy 1993 and in general the Total Quality Control movement). More recent but likewise theoretically disoriented approaches can be found in Mullins, 1996. Available literature on social audits, *le Bilan social de l'Enterprise*, and corporate social responsibility attempt to quantify companies' overall (economic, ecological and social) outputs. Such studies, however, lack an appropriate theoretical approach and fail to integrate complex indexes or to include inputs as a measure of inefficiency, which must ultimately be related to an organisation's outputs.

In short, from the earliest attempts quoted above to the most recent papers of which this author is aware, such as Puig-Junoy (2000), Surruca (2003) or Vergés (2004), which introduced important advances in the definition of the concept, the hypothetical impossibility of the endeavour may still be said to be accepted. The explicit rationale for this hypothesis is based on the lack of a general comparative model able to generate the necessary agreement among experts.

This book, however, attempts to show that such a model exists, subject only to deployment of the effort needed to attain theoretical integration, and thereby to provide the scientific grounds for the above hypothesis (see the Reference Pattern of Values in chapter 3).

Taking these conceptual premises as a point of departure, and applying them to profit-making organisations, the initial question should be re-formulated so as to cover wider but at the same time more specific ground. When can a company be said to be efficient? Initially, as argued above, when it is simultaneously "ecological", "efficacious", "effective" and "incremental". And it must be all these things with respect to the relevant competitors, for nothing can be said to be good/bad, tall/short, ugly/beautiful and so on, unless by reference to some standard of comparison. A company may be highly ecological, efficacious, effective and incremental, but may still be the least ecological, efficacious, effective and incremental of all companies in the same industry and of comparable nature and size. The definition of the new concept calls, then, for the introduction of at least one more dimension: the internal/external relationship that compares company results to those of its relevant competitors. Although this information cannot always be readily gathered, it is becoming increasingly more accessible on the Internet where many or most of the data needed for such assessments can be obtained.

The conceptual model for business efficiency (BE) would, therefore, have to be defined in terms of six dimensions: ecological, efficacious, effective, incremental, profitable and adapted. All of the foregoing is based on the assumption that the set of valid indicators operationalises the theoretical Reference Pattern of Values and Company Stakeholder models. Otherwise, the utility of the approach would have to be challenged or the approach redefined.

But in addition, "organisational efficiency" is viewed from a number of operational perspectives, to enable the reader to compare and decide which approach best fits his or her specific organisation or problem. Initially, the authors of all the chapters of this book agree that to be comprehensible, the definition of efficiency must be couched in terms of inputs and outputs. Furthermore, several applications of the concept are discussed in the various chapters (from small organisations up to the world as an unbalanced social system) for a better grasp of the concept's potential.

This explains why Vojko Potocan begins by introducing a full range of efficiency theories from very different points of view. He then classifies the theoretical approaches under four headings: 1. goal-based approaches; 2. systems approaches; 3. constituency approaches; and 4. contingency approaches.

Potocan's paper revolves around what he calls the "vital attributes" of the concept of efficiency. The first is relativism, i.e., the overall evaluation of organisational effectiveness entails no assumptions about the prevalence of one component over any other. Under the power perspective, an effective organisation satisfies the demand of the most powerful members of the predominant group so as to ensure its ongoing support and the survival of the organisation. From the social justice perspective, based on Rawls' Theory of Justice, an effective organisation minimises the suffering of its most grievously afflicted constituency. From other ethical perspectives, concepts such as utilitarianism, individualism, moral rights, distributive justice, procedural justice, and others are taken into consideration, although the criteria used for evaluation differ. When adopting the evolutionary perspective, the model emphasises the ongoing process of becoming effective rather than being effective, since even the definition changes continually. And finally, from the interest perspective, the emphasis is on viewing the organisation as an association based on the interests of its internal and external participants. All in all, then, Prof. Potocan provides a fairly full overview of the theoretical approaches to the organisational efficiency, stressing the usefulness of systemic postulates.

In his axiological approach, Parra-Luna applies systems theory to the "organisational efficiency" concept, elaborating on Potocan's "interest perspective". Prof. Parra-Luna maintains that any organisation has but one aim: to meet the universal needs of its population, which depend both on the organisation's internal operation and on the conditions prevailing in its surroundings. This approach focuses on the individual and his/her pursuit of some or all of the universal values. Denominated the Referential Pattern of Values (RPV), the list includes health, wealth, security, knowledge, freedom, distributive justice, conservation of nature, quality of activities, moral prestige and power. The first conclusion drawn is that it is inherent in the nature of social organisations to strive to achieve a certain level of one or more of these values, i.e., that they cannot do otherwise. A second conclusion is that this "system of values" can be measured and compared to the "inputs" expended as a preliminary calculation of organisational efficiency, even if that would initially appear to be easier said than done.

In the following chapter, Parra-Luna applies axiological principles to measure organisational efficiency by defining each of its theoretical dimensions through subsystems of objective and subjective empirical indicators. On these grounds, he proposes a new version of the social audit or a new approach to the balanced scorecard for monitoring efficiency. This new tool measures achievements in the following dimensions: ecological; efficacious, effective, incremental, profitable and adapted. All these dimensions are normalised for comparison by means of a general efficiency index (E), where:

if $E>1$ the organisation is efficient; if $E<1$ the organisation is inefficient; and if $E=1$ the organisation is indistinguishable from comparable organisations. Operational and practical solutions are provided for some of the technical problems involved.

Chaime Marcuello deals with the problem of measuring the efficiency of non-profit organisations, whose number is on the rise in today's world. Prof. Marcuello begins by acknowledging that measuring efficiency in such organisations is a fairly fuzzy concept. He nonetheless proposes two main models: in the first, the approach to the various phases of evaluation (context, origin, surroundings, means (human, technological, financial), social networks, participants, communication, plurality, permeability and others) is qualitative. In the second, he studies the possibility of applying the axiological systems theory model. Using Parra-Luna's Referential Pattern of Values, this entails measuring the individual levels of satisfaction achieved, multiplying that value by the number of people concerned, and then dividing the product (the outputs) by the total project cost (the inputs).

Extending the concept of efficiency to specific tasks within an organisation, Matjaz Mulej introduces a method for evaluating official invention-innovation projects. Rather than attempting to measure organisational efficiency as a whole, he measures one of its more important dimensions in modern entrepreneurial life, innovation. He then addresses the various phases of the problem, starting with the problematic definition of "innovation" itself, and paying particular attention to the pre-requisites. These he identifies as: human preconditions, metrics, requisite holism (sustainable enterprise and happy people). For management, innovation is a consequence of the knowledge needed about both concepts: measurement itself and the overall results of the process. But for Prof. Mulej, the measurement of final achievement does not suffice because the requisite holism calls for more viewpoints on synergies and suggests that over-simplification may lead to poor decisions and unsuitable action. To avoid that, he classifies the empirical indicators used into four groups: investment, result, innovation and process. He also develops data sources for resources, capabilities, leadership and process. Finally, he provides a very interesting appendix describing the empirical measurements applied to innovation. This new operational method for measuring innovation is one of the most significant dimensions of organisational efficiency.

Antonio Sanchez-Sucar's article introduces another interesting empirical exercise, the measurement of environmental efficiency in air transport companies. First he poses the question of the feasibility of such an endeavour, given the complexity and chaos that prevail in today's world. To answer this question, Sanchez-Sucar reviews air transport policy and the industry's commitment to sustainability and proposes some indicators with which to measure its efficiency from the Systems Theory perspective. He divides his review into economic, social and environmental efficiency, applied to regulating bodies, service providers (airports), service providers (air navigation systems), technology hardware providers and airlines. He then proposes an emissions efficiency index which could be used to measure the efficiency of procedure generation processes. Special emphasis is placed on the importance of the corporate social "image" in air transport. With this article, Sanchez-Sucar provides a contribution to an approach permitting the measurement of an important dimension of organisational efficiency, namely the "nature conservation" dimension in Parra-Luna's reference pattern of values

Finally, in the last chapter, Prof. Jiménez López critiques the axiological approach applied here, in response to a need identified when the feasibility of the measurements proposed were reviewed. Prof. Jiménez López discusses his well-known theory for protecting the physical environment world-wide to raise the likelihood of survival of what he calls *Homo sapiens sapiens*. He levels radical criticism against the world's socio-political organisation, showing that Western democracy favours a small minority of capital investors and permits the majority to be kept uncomplainingly and submissively working and consuming all manner of goods. He claims that "sustainable development" is mere wishful thinking intended to combine a peaceful and a violent approach. While appearing to be the banner of progress, convenience and scientific technological advancement, development ultimately means obsolescence, for it renders traditional peoples, their cultures, the nature of their habitats and their sense of community obsolete. In this context, according to Jiménez López, any overall measure of organisational efficiency would be impossible to implement.

Nonetheless, one initial reply to this critique is that the axiological approach in systems theory aims to fulfil a dual function. First, it takes a whole set of reference values into consideration. In that context, the survival of *Homo sapiens* is only one of its dimensions. And second, it shows that human beings must not only survive, but also have access to all the values in the reference system, which in the end spur human motivation and action, justifying life in society.

Summing up, the first aim of this book is to present, define and measure a new concept of "organisational efficiency" which is not limited to known economic aspects or related to neoliberal premises or other ideological misconceptions. On the contrary, in the authors' opinion, organisational efficiency must address the entire system of values, projected or attained. Consequently, duly substantiated criticism can and must be levelled against any society's or organisation's system of values. More specifically, the texts hereunder constitute a preliminary attempt to set up an operational quantitative methodology for that purpose.

The second aim is to introduce different approaches to quantifying efficiency applied to specific problems within organisations. On the whole, all the articles identify a number of ways of addressing organisational efficiency, providing a better understanding and critique of this concept.

References

Campbell K.S. (1977) "On the Nature of Organisational Effectiveness". In New Perspectives on Organisational Effectiveness. ed.by P.S. Goodman P.S. & Pennings J.M. San Francisco CA: Jossy Bass.

Dalton R.D. & Kesner I.F. (1985) "Organisational Performance as an Antecedent of inside/outside Chief Executive Succession: An Empirical

Assessment". In Academy of Management. Journal 28 (4).

Ghemawat, P.; Ricart C., Joan E., "The organizational tension between static and dynamic efficiency", IESE, DI-255-E.

Goodman P.S., Atkín R.S. & Schoorman F.D. (1993) "On the Demise of Organisational Effectiveness Studies". In "Organisational Effectiveness: a comparison of multiple models". ed. by Cameron K.S. & Whetten D.A. New York: Academic Press, 163 - 183.

Handy C.B. (1993) Understanding Organisations. London: Penguin.

Hannan M.T. & Freemann J.H. (1997) "The Population Ecology of Organisations". American Journal of Sociology 82, 924-64.

Lazarsfeld P. (1965) "Des Concepts aux indices empiriques". In Le Vocabulaire des sciences sociales, ed. by Raymond Boudon et Paul Lazarsfeld Mouton, Paris.

Mahoney T.A. (1977) "Managerial Perceptions of Organisational Effectiveness". Administrative Science Quarterly 14.

Miles (1980) Miles R.H. (1980) "Macro organizational Behavior. Santa Monica, CA. Goodyear Publishing Company.

Morgan G. (1980) "Paradigms, metaphors and puzzle solving in organisational theory". Administrative Science Quarterly 25.

Mullins J. (1996) "Management and Organisational Behaviour". Pitman Publishing.

Peters T. & Watermann R. (1982) In search of excellence. Warner Books.

Puig-Junoy y Dalmau E. (2000) "¿Qué sabemos acerca de la eficiencia de las organizaciones sanitarias en España?: una revisión de la literatura económica". XX Jornadas de Economía y Salud Palma de Mallorca. 3-5 mayo.

Seashore S.E. & Yuchtman, E. (1967) "Factorial Analysis of organisational performance". Administrative Science. Quarterly 12.

Surruca, J. (2003) "Gobierno de la empresa y eficiencia en organizaciones orientadas a los interesados: una aplicación a las cajas de ahorro y a las cooperativas de Mondragón", Tesis doctoral, U.A. de Barcelona.

Vergés J. (2004) "La eficiencia (productiva) y la relación "principal/agente" en el caso de las Eps. Documento de trabajo.

2

Applying Axiological Systems Theory: the Reference Pattern of Values as an Evaluation Tool

By Francisco Parra-Luna

Abstract

Social Systems (from global societies to small organisations) are made up of human beings. Acknowledgment of this fact has important epistemological implications, since, contrary to some theoretical formulations, (e.g., the "society without men" of N. Luhmann), it is necessary to regard human beings as the crucial element of any type of society in order to perceive their needs and therefore the values supposedly intended to satisfy them. The "need/value" binomial thereby becomes the essential prime material of sociological analysis. Both *needs* as a factor of motivation and *values* as a factor of satisfaction can be operationalised and quantified to depict the principal achievements of complex social organisations. Standardised, and therefore comparable, "axiological profiles" provide a tool that can be generally applied to establish a preliminary measure of the degree of overall organisational efficiency.

But General Systems Theory has not tackled the problem of a generally applicable model of human needs, although mention must be made of the works of Terleckyi, van Gigh, Hall (1994), Buchanan (1999) and others, who have come closest to it. However, the socio-systemic approach as a whole, which could have been expected to focus on the theoretical and ultimately empirical teleology of social systems (the description and measurement of their goals and therefore their progress/regression as a core task) has not necessarily adopted a "sociologising" point of view, that is to say, it does not contemplate the system from the point of view of the man in the street. This has resulted in neglect of the central and ultimate aim of Sociology as a science - collective as opposed to individual efficiency. With a view to merely suggesting an ethical reflection along these lines, I present an axiological approach for calculating the concept of Organisational Efficiency.

1. Introduction

Theories on society have played a relevant role in social research. Much of the present knowledge about societies has been accumulated thanks to these more or less empirically verified sets of ideas (not always organised into systems of propositions). The hypotheses or principles set out below, representing an attempt to introduce an axiological concept of society, are drawn from or inspired by a critical theory of society (itself as object theory and knowledge method) and therefore will (or should) inescapably tend to relate to all other known theories, in the realisation that they all hold some degree of truth. Any theory, focus or method, no matter how innovative or revolutionary it may seem, can only inspire us to complement already existing theories (Flood and Jackson, 1991a, 1991b).

At this time, two new core ideas seem to be permeating the field of sociological theories, partly covering previous confrontations between opposing pairs such as consensus/conflict, empirical/critical theory or grand theory/symbolic interaction, to name a few. These two new core ideas turn on the contrast between Realism and Constructivism on the one hand and Systemism and Postmodernism on the other.

Realism stresses the objectivity of the real world, irrespective of the standpoint of the observer, arguing that observation is adequately and implicitly validated by an orthodox use of the scientific method (intersubjective agreement among experts); whereas Constructivism tends to present "reality" as an essential subjective construction on the part of observer, and hence advocates the inexorable need to "observe observation", sustaining the validity of the practice of what is called "second-order thinking" or "second-order cybernetics". According to the other core idea, systemists see societies as suites of interrelated elements in pursuit of certain ends, where no individual part can be understood without an understanding of the whole, calling therefore for an epistemological vision of the "forest" instead of the individual "trees"; postmodernists, in turn, tend to analyse the fragmented elements of the whole with a ruthlessly critical eye, emphasising, by contrast, the irregularities, the singular, the unexpected....in short, the unexplainable, complexity and chaos. But far from being barren or unprofitable, these apparently contradictory theoretical approaches can be said to contribute much more to an insight into and baring of social structures than if such perspectives and controversies did not exist. Suffice it to recall the conversational applications of Hegelian dialectics (the result of confronting thesis with antithesis, which transcends both), or Ortega y Gasset's perspectivism in his well-known contemplation of the Guadarrama mountains. My personal position is very close to Best and Keller's (1991), summarised by the authors in the following words: "....while it is impossible to produce a fixed and exhaustive knowledge of a constant changing of social processes, it is possible to map the fundamental domains, structures, practices and discourse of society and how they are constituted and interact. Thus, in the rest of this conclusion, we shall argue for a supradisciplinary social theory and a combination of micro- and macro-analysis". In any event, in axiological theory, the approach (which proposes stressing the importance of "values" in social analysis) necessarily assumes a parallel tendency towards the best possible integration and synthesis of theories, standpoints, perspectives, methods and techniques considered to be most fitting at any given time and under any given circumstance.

Moreover, the substantial diversification of sociological theories, the vast number of problems addressed and the numerous specialties, schools and tendencies into which the discipline has branched out make sociology a colourful mosaic of ideas, facts and hypotheses difficult to paste together and vest with any sense of unity, insofar as they are expressions of the socio-human project. Nonetheless, it would seem to be possible to devise a theory which could draw from such accumulated complexity, emerging as the suite of global properties common to the behaviour of individuals, adjusting to the sole possible global structure that arises from the chaos of individualism and inevitably addressing the most common and universal problems facing human beings at the dawn of the twenty-first century; such a theory may attempt to introduce a new (always complementary) avenue for great theoretical understanding, precisely because it focuses on the most essential of human motivations as opposed to the more superficial level of their contingent behaviour. The processes involved in globalisation, communication and the consequential "shrinking" of the planet, with a parallel process of standardisation of needs, desires and fancies, contribute to the heuristic formulation of more universally applicable synthetic theoretical models. I can only agree with Turner (1996) when he says: "In the late twentieth century, it is not too difficult to predict that one possible focus for sociological theory would be the nature of citizenship and human rights...The debate about citizenship has been generated by a concern for the overt decline of governmental commitment to full employment and the welfare state, the changing nature of the state itself, the growth of the global refugee problem, and the increasing ambiguity of the status of children and women in the modern state....It is only by engaging with such political and social issues within the public arena that sociology or general systems theory can hope to survive. Without these political and public commitments, social theory is in danger of becoming an esoteric, elitist, and eccentric interest of marginal academics."

The theories on "values" are not, of course, new in sociology, where at least three prevalent tendencies or schools can be cited: the first approaches the problems of sociology as a science supposedly free of values, an idea the development of which can be traced from Weber's well-known position, through Gouldner's (1971) critique, to papers authored by philosophers such as Rorty (1979), to name only three familiar references. The second regards society as a producer of values, a treatment present in papers ranging from the "Polish peasant..." by Thomas and Znaniecki to recent studies by B. Hall (1994), Buchanan (1991) and M. Hall (1999), not to mention Parsons and the functionalists and, implicitly, all the critical literature that begins with Marx and ends with the postmodernists (Barthes, Baudrillard, Deleuze) including, obviously, the profoundly axiological critiques emanating from the Frankfurt School. The third, which conceives of values as "life systems", can be divided into two branches: the anthropological-cultural (Kluckhohn, Strodtbeck, Morris, Linton....), branch, and the one developed around the concept of business and organisational "culture" (Harrison, Handy, Deal and Kennedy, Schein, Garmendia....).

A review of this literature suffices to acknowledge the theoretical importance of the concept of "value" in sociology, in all of its myriad versions, but its possible theoretical exploitation as an operational "prime material" for social issues does not seem to have been sufficiently developed. The axiological theory that is the object of the present attempted sketch, in pursuit of such operationalisation of the "prime material", falls under the second tendency (production of values), although it overlaps on occasion with the other two, essentially with the Weberian conception of social science. Weber, a critic of and at the same time subscriber to the historicist and utilitarian approaches, to objectivism and subjectivism, represents the earliest methodological effort, which has yet to be surpassed, to make a scientific analysis of society feasible. From the scientific viewpoint, Weberian methodology continues to be a challenge.

With this in mind, it is possible to delve into the nature of individual motivations to action and the pursuit of organisational efficiency.

2. The individual: the key element in any society

Ever since the Greek philosopher Protagoras sustained, seven centuries before the Christian era, that "man is the measure of all things", our view of history and humanity has revolved around the notion of man, the individual, the person or simply the human being, male or female. In ancient times, "man" never ceased to be the centre and ultimate purpose of a universe where even the gods were steeped in human passion. The Greeks, headed by Sophists such as Socrates, were the first to define the human being as rational, logical and social. They raised our species to the highest possible category, but definitely considered it to be subject to a variability that led Plato to say that man can be like a god or the exact opposite, i.e., all or nothing. But it was the Stoics in particular, following in the lead of Zeno of Citium, who claimed that with his power of reasoning, man is able to understand the supreme good, which is to be found in the effort to attain virtue, all else (including pleasure and pain) being irrelevant. This philosophy endured through Roman times and into the early centuries of the present era, with prominent followers such as Seneca and Marcus Aurelius.

Subsequently, despite the creationism introduced by religions such as Judaism, Christianity and Islam (man is created in the image and likeness of God), the Greeks' basic idea underwent further development. Indeed, the human being never ceased to be the absolute protagonist of history for poets such as Dante or Petrarch in Medieval Italy, distinguished French rationalists such as Descartes or the subsequent encyclopaedists such as D'Alambert, Diderot, Rousseau or Voltaire, not to mention reformist Christians such as Erasmus, materialist philosophers such as Marx or modern existentialists such as Heidegger or Sartre. And this humanistic concern has logically informed what are known today as "behaviourist" (G. Ryle's "logical behaviourism", for instance) or etiological (K. Lorenz, for example) theoretical concepts. In short, irrespective of the controversy recently stirred up in the USA between neo-creationists ("intelligent design" theory) and Darwinians (evolution of species), science regards the human being as a part of nature, itself stemming from spontaneous generation in accordance with the laws of chance and necessity (Monod) and the result also of evolution and adaptation to the medium in accordance with evolutionist theories.

And today, the human individual is positioned as the central figure, the origin and the ultimate purpose of the twenty first century company. Given, then, that companies are a compendium of persons and depend on them alone, a review of people's possible roles within the corporate machinery is required, which need not differ widely from the roles assigned them many centuries ago by the Platonic school: The "Theory of Actors-System Dynamics (ASD) of Tom Burns et al. (2001) is a good example of the need to consider the individual person as the main protagonist and as responsible for what social systems are doing.

3. The company's axiological role

Companies, like all other human organisations and societies, are ultimately nothing more than factories that produce values, in the philosophical sense of the word.

In this regard, companies make a much more than satisfactory economic contribution to society, but tend to fail miserably in other areas, a situation that has a negative impact on Social Welfare: i.e., they may be highly successful in terms of most values, but much less so in others. For these reasons, it has become indispensable to take impartial stock, albeit summarily, of what today's company means in axiological terms, both for itself and the surrounding society, in the understanding that no such review is possible without a rigorous analytical method. Such an analytical operation must be preceded by an explicit description of two theoretical models: the pattern of values pursued by all human beings and the stakeholders on whom a company depends.

The "Reference Pattern of Values": defining Social Welfare

If people are the core reference in Social Welfare, the first thing that an entrepreneur must know is what motivates them: what needs, desires or interests induce them to work, cooperate or otherwise maintain relations with his/her company. Once such motivations are identified, all a company needs to do is focus on fulfilling these needs etc., to the extent compatible with its highest possible profitability. Entrepreneurs should, then, take a decisive interest in the world of human motivation, which is tantamount to taking an interest in the values pursued by human beings, in their working and non-working lives.

But participating in the world of values involves coping with vast cultural and anthropological complexities, in which a prior distinction must be drawn to make it possible to address the business world effectively. This basic distinction consists in separating values into two major groups: cultural and universal.

The endless number and variety of the former render them both difficult to handle and of minor significance in the very concrete business environment. The social prevalence of a wide range of behaviours, customs and traditions – amply studied by cultural anthropology – is often limited to the relatively narrow geographic region where they are practised. Scottish kilts, Eskimo nose-kissing, the coin that the Shans of Myanmar place in the mouth of the dead so they can pay for their voyage, or the animistic religions practised in most African countries are but a few examples of behavioural diversity that indisputably enriches human culture. This diversity is, of course, in crisis due to media development and its well-known corollary, a "shrinking" planet, but even so, the endurance and force of these traditions and customs are still hugely – and at times militantly – powerful. The appearance of nationalisms that claim differential cultural traits as a basis for political independence or other demands are proof of their relative permanence. But these are not generally values of direct interest to companies, although they may ultimately also affect them.

The latter group, however, comprises a much shorter list and is identical for the entire human race. It can be sustained that the vast majority of human beings desires and pursues the same or similar objectives in terms of vital or cultural goals that are basic to and inherent in people everywhere. All across the planet, with scant exceptions of little statistical significance, peoples seek to be in good physical and mental health, to feel as secure as possible in the face of feared or unforeseeable contingencies, to attain a sufficient standard of living, to understand unknown and mysterious phenomena to satisfy the species' innate curiosity, to enjoy freedom of movement and expression, to perceive a certain degree of distributive justice and equality of rights among peers, to feel integrated, loved and admired by others in their immediate social circle, to live in harmony with nature, of which the species inevitably forms a part, and finally, in more advanced societies, to feel capable of self-fulfilment and personal growth in the use of their freedom and imaginative and creative aptitudes. Table 1 shows what the author has christened a "Universal pattern of values", which will be used at different points of this study; its theoretical grounds lie in natural law, the theories on human needs and the UN's Universal Declaration of Human Rights, among others.

Indeed, in reply to those who deny the possibility of establishing a universal list of rights applicable to all people everywhere, there are at least three sources in literature from which a valid general theoretical model of both the concept of social efficiency in its broadest sense and of organisational and business efficiency may be formulated for comparison across space and time. These are: 1) the *philosophical* approach (such as in Carr E.H. et al, 1973); 2) the *sociological* approach (such as in Bauer et al., 1966; and 3) the *political* approach (UN; 1948). Irrespective of the philosophical and sociological approaches, which would in themselves suffice as scientific support, this general model is validated by political literature.

The core argument is as follows: the general validity of the United Nations' 1948 Universal Declaration of Human Rights can hardly be denied and indeed it has been internationally reiterated in numerous world-wide conferences: Rio de Janeiro in 1992 (Environment); Vienna in 1993 (Human Rights); El Cairo in 1994 (Demography); Copenhagen in 1995 (Poverty and Social Development); Beijing in 1995 (Women); Istanbul in 1997 (Urban Growth); Seattle in 1999 (World Trade) and Milan in 2003 (again Environment), to name but a few of universal relevance. Such unanimous agreement on human needs of people of all kinds, anywhere on Earth, makes it possible to structure them under a general model. "What political and social organisation generates" – writes Lipovetsky (1992) in his "Le crépuscule du devoir" – "is an obligation not to the divine legislator, but rather to inalienable individual rights... According to these new democratic Stone Tablets, only inalienable human rights are explicitly formulated in terms of the duties stemming therefrom and as obligations that must be inexcusably respected."

An analysis of the content of such declarations, formally accepted by the governments of the planet's 190 countries on behalf of its nearly seven billion people, would yield a model comprising nine basic needs or rights: right to health; economic development; safety; freedom; education and knowledge; equal opportunities; environmental conservation; support in the event of need; and self-fulfilment and personal development.

In research, such rights play the role of the "Reference Pattern of Values" (C. Kluckhohn, 1951) as the sole entities able to meet universal human needs, and correspond to validated theoretical schemes (such as Maslow, 1970). "Value", according to the former author, is the flip side of "need", and together, the two represent the motivation that drives the world. This model can be validated, then, as the basic general model for comparisons across space and time, by contrast to the world's vast, rich, complex and respectable anthropological-cultural variety. The latter, in any event, is not incompatible with the surfacing and satisfaction of the more basic rights or values common to the human species as a whole. For all the foregoing, this author tends to disagree with theorists who deny the possibility of quantifying the concept of "organisational efficiency" for a lack of consensus on the aims and objectives to be pursued.

Consequently, the argument that these rights are interpreted differently by each individual (even though we know they are) is no longer relevant to this discussion. And this is so simply because the general idea underlying each of these rights has been the object of prior universal consensus by those individuals' representatives, to reach a core meaning or "averaged" ideal. These and no others are the value objectives (via the empirical indicators that represent each) whereby UN policy and its agencies (UNESCO, WHO, FAO and so on) are governed, respecting both the redistribution of wealth (aid programmes for the neediest countries) and budgetary issues (what each country should pay to sustain the UN). And these and no others are the values around which governments are typically structured and subdivided into ministries or departments in nearly all the world's countries: from health to economic development, political freedoms, and distribution of wealth.

An entirely different question is the ultimate definition of the number and content of such universal rights or "value objectives" – when it comes to their implementation (on the grounds of objective and subjective indicators) – in terms of: a) the units (countries, companies...) to be compared; b) the period of time involved; c) the purposes of the comparison; and d) the material resources available. Hence, Table 1 includes, for instance, a new "value objective": power as applied to the company, which appears as a pursuable and at times indispensable aim (company consolidation) for business survival. For a discussion of the isomorphic patterns of business expansion in several countries based on the power value, see Kim et al. (2002).

Table 1: The Reference Pattern of Values

No	Need	Value pursued	Symbol
1	Physical and mental well-being	Health (S)	Y_1
2	Material sufficiency	Material wealth (RM)	Y_2
3	Protection against contingencies	Security (Se)	Y_3
4	Freedom of movement and thought	Freedom (L)	Y_4
5	Understanding of and command over Nature	Knowledge (C)	Y_5
6	Equity	Distributive justice (JD)	Y_6
7	Harmony with Nature	Environmental Conserv. (CN)	Y_7
8	Self-fulfillment	Quality of Activities (CA)	Y_8
9	Social esteem	Prestige (Pr)	Y_9
10	Influence	Power (P)	Y_{10}

It should be stressed once again that the applicability of this table lies first of all in that the values pursued are universal in space and time, which means that there is no known society where such values are not in place (although the importance attached to each may differ depending on circumstances). And given the deterministic power of human nature in the pursuit of quality of individual life, there is every reason to believe that these values will extend into the future. Secondly, this model includes all existing values; in other words, any human need, desire or interest should be covered by some one of the ten values in the model, from the need to be physically and mentally fit (Health) to the desire to be immensely wealthy (Material Wealth), including the simple desire to breathe the freshest possible air (Environmental Conservation) or the need to believe in a just God able to provide security against whatsoever contingency may arise in this life (Security). If this were not true in all possible applications, the model would not be valid and would have to be modified. Hence, the operational definitions of the values must include (objective and subjective) empirical indicators to confer scientific validity upon each value concept in terms of time, place and other circumstances.

Attention should also be drawn to the existence of positive inter-value systemic relationships, whereby an increase in any one of them leads to parallel although proportionally different increases in the others. If a population's health improves, for instance, enhancements should logically be expected in its level of education, standard of living, desire for freedom, concern for nature and so on; and vice-versa: improvements in any of these values should have a beneficial effect on people's health.

There is one notable exception to this rule: the dialectic relationship between Freedom and Distributive Justice is clearly indirect, for empirical evidence shows that the greater the degree of freedom, the more unfair is the distribution of goods and services produced. And conversely, absolute equity in distribution cannot be attained without radically curtailing personal liberties. The ocean depths (reign of absolute freedom), where big fish systematically gobble up the smaller fry (reign of absolute injustice) provide one of the clearest examples of this indirect relationship. Consequently, the most highly developed political systems (such as in the Scandinavian countries) have attempted and possibly achieved the best possible trade-off between the two values.

Each social institution, in turn, pursues and maintains a certain characteristic type or system of values out of tradition or need. Such systems consist in the different relative emphasis placed on certain values at the expense of others. While religious institutions stress Security, for instance, education systems stress Knowledge or organisations such as Amnesty International Freedom stress Freedom, companies emphasize Material Wealth (obtaining the highest possible return on the capital invested). Nonetheless, all these institutions, inasmuch as they are social organisations, should also inevitably meet certain standards in the other nine values listed in Table 1, for otherwise the system risks extinction. Be it said, then, although only for the sake of the hypothetical ideal, that the system of values typically pursued by companies is more or less as illustrated by the scheme in Figure 1. Note, in addition to the nearly absolute emphasis on Material Wealth, the relative importance attached to values such as Freedom (business traditionally attempts to minimise political, economic or labour-law constraints) and Knowledge (of growing importance in the information age). Conversely, for different reasons, including their characteristic propensity for economic liberalism or their need to survive in a hugely competitive marketplace ("the survival of the fittest"), they typically show little concern for raising the degree of Distributive Justice. It is not that entrepreneurs object to this value; rather they find it incompatible with the levels of freedom they need to conduct their business.

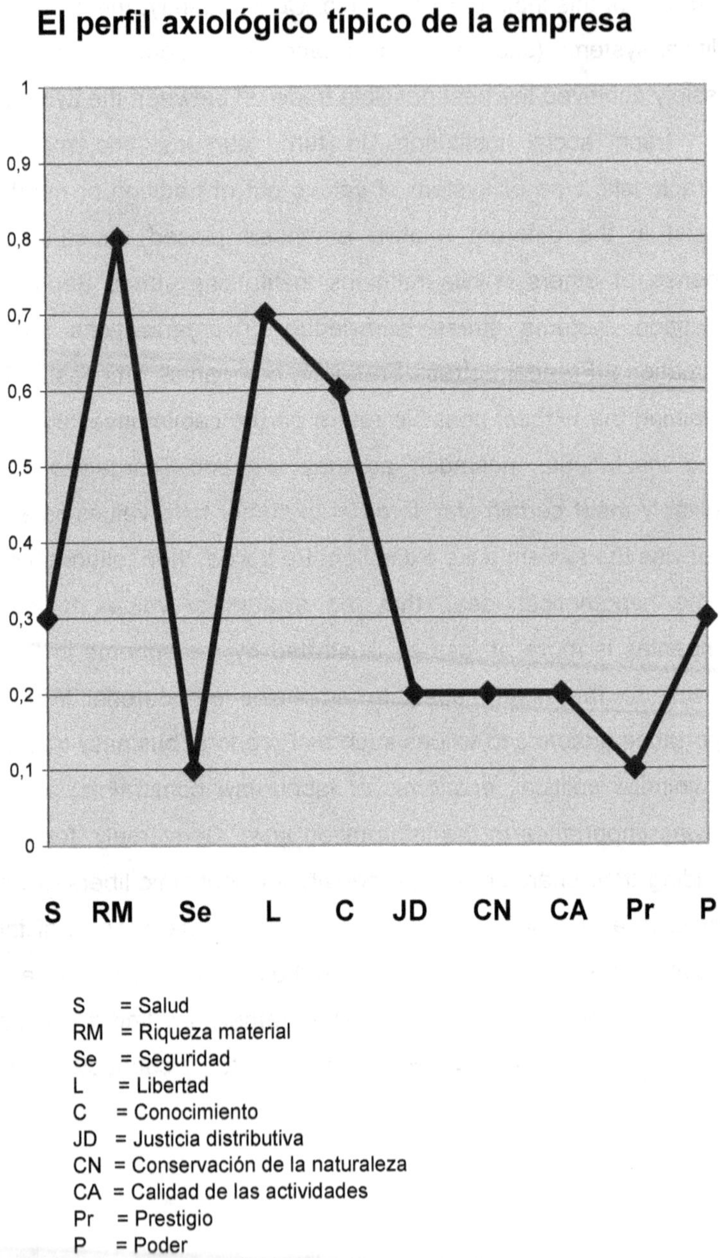

Figure 1: A typical axiological profile for a business concern

Translation	
Spanish	**English**
Salud	Health
Riqueza material	Material wealth
Seguridad	Security
Conocimiento	Knowledge
Justicia distributive	Distributive justice
Conservación de la naturaleza	Environmental conservation
Calidad de las actividades	Quality of activities
Prestigio	Prestige
Poder	Power

This would be approximately the corporate profile perceived by the public at large: an institution focusing on Material Wealth, Freedom and Knowledge, and much less on the remaining values. As a result, the corporate world's actual contribution to Social Welfare is not mirrored in the level of social prestige it is accorded.

Just as companies can be axiologically represented by Figure 1, the system of values pursued by each individual involved with a company can be similarly plotted as his or her personal axiological profile. From this vantage point, people can be viewed as "bipedal axiological profiles" capable, however, of changing with circumstances. This is a logical conclusion to be drawn from the fact that the so-called Reference Pattern of Values in Table 1 is based on the natural and universal essence of human needs. The inference, then, is that no one, regardless of his/her circumstances, is exempt from the motivation associated with one or several values, nor able to avoid the usually unconscious attempt to balance the benefits stemming from the ten values against one another. There is barely an everyday deed or action that escapes this axiological dialogue, be it the decision to quit smoking (Health and Material Wealth), to paint a picture (Quality of activities), or to buy a high-priced automobile. By way of illustration, a person deciding to buy an expensive car, for instance, would mentally calculate the trade-off between Prestige and Material Wealth, concluding that there was more to be gained from the numerator (respect from suppliers or neighbourhood admiration) than the denominator (saving the cost). And even in the most drastic situations, when a person deems that life is no longer worthwhile, he/she may possibly be weighing values such as Freedom and Quality of Activities. Finally, since opinions on the relevance of the different values change frequently, the above definition could be restated more precisely by saying that people are, by nature, "bipedal and variable axiological profiles".

In short, whereas value systems (profiles) change, the Reference Pattern can be applied generally across time and space and its values are undeniable, for they are imprinted in human nature. Drawing an analogy to deoxyribonucleic acid, which contains the genetic code that enables living organisms to generate other beings similar to themselves, the RPV (Reference Pattern of Values) would represent the axiological information that enables human beings to aspire, in whatever situation, to the same type of universal values. In Lipovetsky's (1994) words, the RPV would be the "genetic code of modern democracies", more or less the human species' social DNA.

4. The company as a transforming system

The principle of "transformation" in enterprise has always been readily comprehensible, particularly in manufacturing companies, with their incoming raw materials and outgoing finished product. More strictly speaking, however, companies of all kinds have only two types of inputs: capital and labour. And what has always been regarded to be company output (whether material products or services) can no longer be validly viewed as such. Not even the basic and essential objective, which is none other than a suitable return on the capital invested, can be validly regarded to be an output. Because what any company, regardless of type, size or circumstance, produces is always an inevitably complex value system. One such value, indisputably the most essential for private companies, consists in attaining at least a sufficient financial return to ensure their durability.

The foregoing underlies the transformation principle that assumes that Inputs (X) undergo transformation (T) to produce the Outputs (Y) described above in the Reference Pattern of Values. This principle provides a definition of the most direct and simple measure of company "efficacy", which, expressed as T=Y/X, provides for an analysis in terms of the formal aspects of general Social System Theory.

This expression is so universal that not only is a company's existence justified if Y/X is greater than one (standardised output is greater than standardised input), but even the people who work for or with it do so exclusively to obtain the highest possible Y/X ratio. The sole difference is that their ratio is calculated on an individual scale, measuring what the company furnishes them (their personal "Y") in return for their effort in time or energy (their personal "X). In other words, neither companies nor individuals are able to escape the effects of this ratio in real life situations, and this indispensable principle should serve as a guiding star in the conduct of company business.

Indeed, the utility of the expression is such that it is applied more or less unconsciously with every step people take. Hence, for instance, when someone decides to have a beer, he/she has first to decide whether the ratio between satisfaction (Y) and cost (X) is greater than one. That is to say, he/she must find "Y" (satisfaction) to be worth more than "X" (cost). And much the same process is followed by students registering for a course: they value the knowledge to be acquired or the prestige of a degree more highly than the time and money invested; or by the author of the present paper, who regards the pleasure derived to be worth more than the time devoted to writing it. Rare is the everyday circumstance where the ratio T=Y/X is not invoked, where people would not ultimately seek to maximize the value of "T" in circumstances in which they deploy some effort.

An even fuller understanding of Social Systems Theory can be gleaned from a comparison to Physical-Natural Systems Theory. Whereas in natural systems the Y/X ratio seldom reaches one (outputs cannot be greater than inputs, inasmuch as both are predetermined), this changes substantially in social systems or systems involving human intervention. Indeed, in physical-natural systems the Input-Output transformation may give rise to admirable and extraordinarily beautiful change, such as when certain earth minerals plus air and water combine to create something as lovely as a flower; or when the most primary elements of life unite in aquatic worlds to generate fish of surprising beauty and colour. But in these cases, developments follow a perfectly predetermined programme designed by nature which changes either not at all or so slowly (evolution) that the transformation cannot be directly observed. To put it another way, in physical-natural systems, the outputs are either a function of inputs (Y=f(X)) or if "T" is also involved (Y=f(X,T)), "T" is predetermined by a natural programme that is unchangeable unless human beings intervene, such as in plant grafts or the genetic modification of plants and animals.

But social systems, and certain man-made mechanical systems, are totally different. Here, the explanatory importance of the Outputs (Y) lies neither in the resources (X) nor in any type of invariable programme defining (T); rather, this third element of the equation changes instantaneously and constantly due to human action (information and knowledge) in the wake of contingencies deriving from social, technological or any other manner of change. That is to say, the intrinsic importance of "T" lies in its constant self-criticism, its self-consideration as a consistently underdeveloped, disorganized system and the ability to change its design in pursuit of the most efficient Y/X ratio possible. This can be achieved because "T" depends not on a set or predetermined programme, but on the ingenuity, will, creativity or imagination of the people comprising the respective social system. While in physical-natural systems, by virtue of their changeless transforming programmes, two plus two is four, in social systems, by virtue of their continually renewed transformation programming, two plus two may be four, but it may also be twenty, thirty, one hundred or one thousand, or less than four, or even be less than zero when they malfunction.

A social system may multiply its inputs, such as when a happy couple sees its wishes fulfilled, having and successfully raising children; or reduce them, such as when a marriage ends in divorce, shattering the couple's expectations. The outcome depends on the organisation of the transforming elements that human beings are able to assemble in the family system. In business terms, the factors involved include the original means-end design, organisational structure, critical mass in respect of the market, the best possible staff selection, personnel placement in terms of present capacity and but also of potential, lifelong learning, poise, ambition, social relations and so forth, all in terms of that ubiquitous production factor that today goes by the name of "information". Without such personal and intangible assets, there can be no company. And if God rested after programming nature, that same God must have afforded humanity the freedom to recreate and reinvent its own world, making us, in turn, gods in His image and likeness, passing on His responsibility to such an extent that we and we alone, the members of the human race, are the sole makers able to set up and profitably run a modern enterprise.

Coming back to a premise established earlier, companies can be reduced to just two components: capital and labour. But a distinction should be drawn between the two: despite the unquestionable importance attached to capital and saving since the appearance of the first large companies in the eighteenth and nineteenth centuries, or more explicitly, since the advent of Weber's theories that identified the Calvinist roots of capitalism, capital is not, nor can it be, as important as labour. For what is capital, ultimately, if not work or effort, or savings accumulated from work? All or nearly all the things that make up a company consist in or are the outcome of labour. Work and effort are what it takes to generate savings, design a new company, enlist the necessary material and personal resources (including loans, with no initial capital at all), establish social relations, plan new development programmes, organize the different parts of a company... In short, companies are created and maintained thanks to – primarily intellectual – work and work alone. Consequently, as is often acknowledged (although rarely with any practical consequences), a company's most important asset is its staff, its personnel, from the Chairman of the Board down to the lowliest employee. And when management focuses on people, the company is limited in what it ultimately becomes only to the extent that its people are limited. Quite obviously, companies are continually faced with financial, commercial, technical, personnel or other types of problems. But the solution to such problems is always or nearly always provided by the people in the respective areas, because only people with their information, experience, effort, imagination, poise or personal networking can solve such problems.

Hence, the typologies by which corporate cultures are classified, depending on whether they focus on cost, quality, people, technology, safety or final profitability, to name a few (Garmendia, 1994), cannot be said to be erroneous. It is probably not a bad idea for each company to portray itself as specializing in one of these cultural traits, using it as a banner or symbol to differentiate itself from others. But this should always be done on the understanding that two hallowed principles must be respected internally: first, that business activity must focus on the full range of cultural traits, without neglecting a single one; and second, that despite this initial equity of "accentuation", one such cultural trait stands out above all the rest: namely, the concern for people, since they and they alone are going to provide the solutions to business problems.

Finally, the T=Y/X concept needs to be extended to each and every one of the operational areas of companies, as if they were separate lines of business. Think of the potential in considering every division, unit, sub-unit, centre, department or section as a system in itself, the sole purpose of which is to obtain an output (Y) from an input (X) and whose primary obligation is to attempt to continually raise the value of the transformation ratio T=Y/X. All relatively independent departments, centres and sections have a function, "Y", that they should maximize (for instance, selling units of product) by drawing on a series of resources, "X" (premises, personnel, overhead...), which must be minimised. There is nothing simpler, then, than to monitor the number of units of "Y" that can be "produced" with the smallest possible quantity of "X", generally cost. Naturally, the periodic quantification of " I ", which translates into system "transforming intelligence", should be the basis for any productivity-based incentive plan. It is assumed that failure to adopt this principle in Spanish companies will entail a loss of "social welfare" in the light of the direct relationship between this concept and "business efficiency".

Conclusion

The need to apply this principle is very nearly universal, from government (in particular) down to all other types of organisation (non-governmental organizations - NGOs for instance); all bureaucratic centres with a purpose to realise and a person in charge should be governed essentially in accordance with the periodic quantification and ongoing improvement of the "T" ratio. If this were accomplished, the world would be a different place and its social welfare enhanced. To begin with, it would be more ecological and sustainable to reduce the consumption of "X" per unit of "Y" produced; and "Y" is a complex value system the maximum success of which consists in reaching its goals in a balanced manner, duly adapted to the prevailing circumstances.

References

Bauer R. et al. (1966) *Social Indicators: a First Approximation.* Cambridge, Massachusetts: MIT Press.

Best S. and Keller D. (1991) *Postmodern Theory: Critical Interrogations.* Hong Kong: Macmillan.

Buchanan B. (1999) "The role of values in measuring performance of social systems". In *The Performance of Social Systems.* ed.by Parra-Luna F. New York: Plenum, 25-36.

Burns, T. et al. (2001) "The Theory of Actors-System Dynamics:Human Agency, Rule Systems and Cultural Evolution". In *Systems Sciences and Cybernetics. ed.by Parra-Luna F.* International Encyclopedia of Life Support Systems.

Carr E.H. et al. (1973) *Los derechos del hombre.* Barcelona: Laia.

Flood RL. & Jackson MC. (1991a) *Critical Systems Thinking: Directed Readings.* Chichester: Wiley.

Flood RL. & Jackson MC. (1991) *Creative Problem Solving: Total Systems Intervention.* Chichester: Wiley.

Gouldner A. (1971) *The Coming Crisis of Western Sociology.* London: Heinemann.

Hall B. (1994) *Values Shift.* Rockport Massachusetts: Twin Lights.

Hall M. (1999) "Systems Thinking and human values: towards understanding the performance of social systems". In *The Performance of Social Systems.* ed.by Parra-Luna F. New York: Plenum, 15-25.

Kluckhohn, C. (1951) "Los valores y las orientaciones de valor en la teoría de la acción". In ed.by Parsons T. y Shils E. *Teoría de la acción social.*

Lipovetsky G. (1994) *El crepúsculo del deber.* Barcelona, Anagrama.

Maslow, A. (1979). *Motivation and Personality.* Harper & Row.

Parra Luna F. (1983) *Elementos para una teoría formal del sistema social.* Complutense.

Parra Luna F. (2001) "An Axiological Systems Theory:Some basic Hypotheses". In *Systems Research and Behavioral Science.* Syst. Res 00, 1-16.

Parra Luna F. (2008) "A Score Card for Ethical Decision Making". In *Systems Research and Behavioral Science.* Syst. Res. 28, 249-270.

Richardson HA. et al. (2002) "Does decentralization make a difference for the organization?". In *Journal of Management.* 28 (2), 217-244.

Rus A. & Iglic H. (2005) "Trust, Governance and Performance". In *International Sociology.* 20(3), 371-391.

Verges J. (2004) *La eficiencia productiva y la relación principal/agente en el caso de las EPS.* Documento de trabajo.

Wittmann WW. & Hattrup K. (2004) "The Relationship between Performance in Dynamic Systems and Intelligence". In Systems *Research and Behavioral Science.* 21(4), 393-409.

3

A Numerical Application: The Company as a Case Study

Francisco Parra-Luna

Abstract

Taking the conceptual grounds of the previous article as a point of departure, the initial question would have to be re-formulated in a more general and concrete manner: When is a company efficient? Initially, as it will be argued below, when it is simultaneously "Ecological", "Efficacious", "Effective" and "Incremental". And it must be all these things with respect to relevant competitors, for nothing can be said to be good/bad, tall/short, ugly/beautiful and so on except in comparison to some reference. A company may be highly ecological, efficacious, effective and incremental, and yet still be the least ecological, efficacious, effective and incremental of all companies in the same industry and of comparable size. The definition of the new concept calls, then, for the introduction of at least one more dimension: internal/external that compares company results to those of its (relevant) competitors. These principles should be applied also to public and non-profit organisations.

1. Introduction

Any company concerned about its overall efficiency and aware of the complexities of an age in which technology, information and knowledge prevail, can hardly ignore the existence of balanced scoreboards, tools that measure to what extent it reaches its objectives. This instrument is so essential that there is scarcely a company of any prominence without it, but that does not necessarily mean that it meets all the necessary and possible requirements of the information age.

It is generally acknowledged that, like the notions of industrial and commercial companies, the traditional balance sheets and income statements of classic accounting schemes have grown obsolete. Businesspeople and analysts realized decades ago that the standard financial statements presented by companies at year-end failed to provide the information needed to judge a company's actual value or the quality of its governance. The reason, among others, was that they did not reflect such important aspects as the social (employee attitude) or intellectual (expertise or know-how) dimensions of the organisation. The history of the Social Audit since the 1970s in Europe and the US reflects this change. Two governments (France in 1977 and Portugal in 1985) even made the Social Audit mandatory for all companies above a certain employee size. In Spain, the first organisation to implement social audits was the former INI (National Industry Institute), followed by financial institutions and corporations (Fundación Rumasa, Banco de Bilbao, Caja Madrid and so on). The initial interest has since waned, however.

A number of papers has recently appeared, in the US particularly, revitalising the "Social Audit" concept under the term "Balanced Scoreboard", reflecting the growing conviction that moral or intellectual capital and assets should not be invisible in company reporting. The tool is, of course, also intended as an aid for strategic management.

Despite their recent publication, however, these papers still lack both a systematic theoretical approach and operational itemisation. On the one hand, they do not build from an overall axiological theory covering company stakeholder needs; and on the other, they fail to separate objectives from objective achievements. Finally, no provision is made for strictly comparable standardisation of information. Nonetheless, the approaches adopted initially by Kaplan and Norton (1997) and followed in several papers authored by a consultant firm, Horvath & Partners, and summarised in Horvath & Partners

(2003), focus on reducing the scoreboard indicators to the most strategic items and on intensifying management and control as needed to reach the goals proposed. What they advocate is a strategic rather than a comprehensive Balanced Scoreboard (not all the indicators are included).

Other interesting tendencies address concepts such as "Corporate Social Responsibility" or the measure of the various forms and denominations of a company's intangible assets (intellectual, human, social, relational, structural, organisational, technological, customer-oriented and so forth). Some of these will be mentioned below in connection with the calculation of the "Corporate Intelligence Quotient" (CIQ) and many have been the object of reports by companies of the prominence of Telefónica, Colgate-Palmolive, Eisai Co. Ltd, Samsung and World Bank, to name a few.

The methodology adopted on the occasion of the present study, termed Corporate Balanced Scoreboard (CBS), draws from the various methodologies proposed by other authors. The present approach attempts to include the characteristics listed below, some of which embody some degree of theoretical or operational added value with respect to prior formulations.

1) The theoretical basis for this scoreboard is the Reference Pattern of Values and the corporate stakeholders; in other words, it corresponds to systemic wholes.
2) It covers both social and economic aspects.
3) It uses both objective (statistically recorded facts) and subjective (quantified record of opinions) data.
4) It standardises data in coefficients that fluctuate around the number one, facilitating interpretation and comparison as well as integration into more complex indices.
5) It includes standard management control through routine "forecast – follow through - deviation" procedures.
6) It subdivides the results by department or area of responsibility, as well as by Reference Pattern values and their component indicators.
7) Since standardised indicators are used, the results can be charted on

graphs to render the evaluation process more efficient.

This CBS pursues three primary objectives:

1) To serve as an aid for the best possible diagnosis of a company's situation, showing its strengths and weaknesses with a view to identifying suitable remedies.
2) To provide for strategic planning based on a second selection and discussion of the indicators regarded to be most decisive for the company at any given time.
3) To establish integrated and standardised management control with comparable indices.

The stages of periodic CBS formulation (monthly, semi-annually, annually) are much the same as in conventional management control, namely:

(1) Preparation

- (a) Establishment of the general outline of company policy for each financial period considered in the context of the company's long-term mission.
- (b) Formulation of a list of (objective and subjective) indicators that translate the policy defined in the preceding item.
- (c) Establishment of the scores to be attained in each of the indicators on the basis of the previous period's results and of the objectives for the following period.

(2) Follow-through

- (d) Performance of the necessary actions to reach the proposed level.
- (e) Analysis of the deviations at the end of each period, presenting a chart classifying efficiency levels by department or area of responsibility.
- (f) Routine meetings with the various department or area managers to correct possible deviations, after analysing the aetiology or cause.

In short, the primary difference between this CBS and the former "Social Reports", "Social Audits" "Corporate Social Responsibility Reports" and similar lies in the fact that the CBS is an analytical tool to be used for in-house or strategic purposes rather than for public reporting.

2. Indicators to be used: an initial suggestion

Some of the indicators that may be appropriate for a comprehensive CBS in a medium-sized/large industrial company are to be found in the principles underlying the Value/Stakeholder Matrix. A standard organisational chart includes company General Management (G), Human Resources Department (HR), Commercial Department '(C), Production Department (P) and Administration (A) (listed here in random order).

The methodology for selecting social indicators depends on a series of theoretical and methodological requisites (at least twenty) which are consolidated here based on the following criteria: they must be 1) relevant (theoretically significant); 2) reliable (verifiable by means of quantitative data that reflect reality); and 3) assumable (adapted to the company's size and financial capacity).

The indicators are presented within their own units to permit subsequent calculation of forecast and actual improvement or regression, to provide information on both the deviation in percentage computed from absolute figures and the ratio between the two periods (previous and current) in the form of indices.

The CBS subdivides indicators into two major groups: Outputs and Inputs. The former vary as widely as company objectives; the latter are characterised by a single expression: the economic outlay budgeted/affected to produce the outputs. The ultimate purpose of the CBS is to compute the T=Y/X ratio at all possible levels (overall and by area of responsibility).

Reference Pattern Values

Each of the Reference Pattern Values is measured in term of the indicators set out in the chart below. The meaning of the characters in parentheses following each indicator is:

a) The first parenthesis denotes whether the indicator is positive or negative, i.e., whether good management should pursue their increase (+) or decrease (-).
b) The second parenthesis contains three numbers separated by commas: the first denotes achievement in absolute values in the recent past, the second the percentage of improvement over the preceding period and the third achievement in absolute values at the end of the present period.
c) The third parenthesis specifies the relative weight of each indicator, on a scale of 1 to 10.
d) The fourth parenthesis indicates the respective area of responsibility.

Hence, indicator 1.1, the first on the list, titled "Absenteeism due to illness in days" is signed negatively (-), meaning that growth is undesirable; 215 days were lost in the recent past; a 5% improvement (decline in the number) was foreseen; actual absenteeism came to 217 days; the relative weight or importance on a scale of 1 to 10 is "5"; and finally responsibility is attributed directly to the Human Resources Department (HR). The following is a set of example value for an output Y and input X Reference Pattern.

Outputs (Y)

1) Health

Absenteeism due to illness, days (-)(215,5,217)(5)(HR)
Percentage of staff undergoing a yearly physical examination (+)(HR)
Employee opinion on "occupational health" (+)(8,5,8.11)(1)(HR)

2) Material Wealth

Gross profit (+)(G)
Productivity (+)(120,6,125)(9)(G)
Average salary (+)(G)
Average commission per agent (+)(C)

Dividends (+)(G)

Employee opinion of their total income (+)(7.7,5,10)(1)(HR)

3) Security

Allocation to reserves (+)(G)(20,5,21)(4)(G)

Accidents (-)(P)

Fixed asset depreciation at replacement cost (+)(P)

Supplier diversification (+)(P)

Diversification of customer base (+)(C)

Diversification of agents (+)(C)

Risk coverage (+)(G)

Short-term (30-day) liquidity (+)(A)

Medium-term (90-day) liquidity (+)(A)

Capital structure (equity/borrowed capital)(+)(A)

Staff severance (-)(HR)

Average retirement pension (+)(HR)

Average staff seniority (+)(HR)

Employee expectations for the future (+)(HR)

Shareholder expectations for the future (+)(G)

Supplier expectations for the future (+)(P)

Agent expectations for the future (+) (C)

Customer expectations for the future (+) (C)

Percentage of permanent staff over total headcount (+)(HR)

Employee opinion of their level of security (+)(6,5,5)(10)(G)

4) Freedom

Staff information meetings (+)(12,0,15)(1)(HR)

General employee surveys (+)(HR)

Specific employee surveys (+)(HR)

Supplier surveys (+)(P)

Agent surveys (+) (C)

Customer surveys (+) (C)

Participation in trade union elections (+)(HR)

Union membership rate (+)(HR)

Employee opinion of their degree of freedom in the company (+)(7,5,7)(6)(HR)

5) Knowledge

Percentage of staff with full university degrees (+)(25,10,35)(2)(HR)

Percentage of management staff (+)(G)

Percentage of time devoted to product development (+)(P)

Percentage of time spent in electronic data exchange (+)(G)

Proportion of products on the market for less than two years (+)(P)

Percentage of R&D spending on basic research (+)(P)

Percentage of R&D spending on product design (+)(P)

Percentage of R&D spending on applications (+)(P)

Investment in new product support and training (+)(P)

Average age of company patents (-)(P)

Percentage of employee suggestions accepted (+)HR)

Percentage of partnering suppliers (+)(P)

Percentage of employees participating in training courses (+)(HR)

Percentage of employees involved in mentoring (+)(HR)

Number of companies networked (+)(G)

Employee opinion of level of knowledge (+)(HR)

Subjective degree of Business Intelligence (+)(8,10,8)(8)(G)

Corporate Intelligence Quotient (CIQ)(+)G)

6) Distributive justice

Percentage of net profit allocated to reserves (+)(G)

Ditto, dividends (+)(G)

Ditto, employees (+)(37,5,38)(7)(HR)

Income inequality rate (-)(HR)

Gender income inequality rate (-)(HR)

Percentage of capital held by employees (+)(HR)

Percentage of employees holding company shares (+)(HR)

Percentage of employee representatives on the board (+)(HR)

Percentage of women managers (+)(HR)

Overtime agglutination (-)(HR)

Night-time or arduous work agglutination (-)HR)

Part-time work agglutination (-)(HR))

Flexible schedule agglutination (-)(HR)

Employee opinion of equity in the company (+)(6,10,6)(6)(HR)

Supplier opinion of equitable treatment from the company (+)(P)

Agent opinion of equitable treatment from the company (+) (C)

Customer opinion of equitable treatment from the company (+) (C)

7) Environmental conservation

Percentage of landscaped surface (+)(G)

Noise level (-)(P)

Air pollution (-)(P)

Pollutant waste emissions or spills (-)(15,5,10)(2)(P)

Fines for pollution (-)(P)

Employee opinion of in-house pollution (-)(P)

Opinion of surrounding community regarding pollution emissions or spills (+)(7,5,7)(2)(P)

8) Quality of activities

Suggestions submitted by employees (+)(300,5,320)(7)(HR)

Suggestions submitted by suppliers (+)(P)

Suggestions submitted by agents (+) (C)

Suggestions submitted by customers (+) (C)

Average working hours per week (-)(HR)

Percentage of employees with flexible hours (+)(HR)

Percentage of part-time employees (+)(HR)

Percentage of telecommuters (+)(HR)

Percentage of employees working overtime (-)(HR)

Percentage of employees working nights or doing arduous work (-)(HR)

Percentage of employees rotating in their jobs (+)(HR)

Decision-making at lower hierarchical levels (+)(HR)

Percentage of employees participating in management control (+)(HR)

Proportion of group incentives over total (+)(HR)

Social-workplace climate (+)(6,10,6)(8)(HR)

9) Prestige

Unpaid bills (-)(A)

Fines (-)(G)

Percentage of spending on donations, sponsorship and philanthropy (-) (C)

Percentage of in-house spending on good works (+)(HR)

Company prizes or awards (+)(G)

Customer complaints or returns (-)(38,5,30)(1) (C)

Supplier complaints (-)(P)

Employee complaints (-)(HR)

Agent complaints (-) (C)

Exports (% of turnover)(+) (C)

Number of export host countries (+) (C)

Franchises granted (% of turnover)(+) (C)

Legal claims against the company (-)(G)

Employee opinion of company prestige (+)(6,4,5)(10)(HR)

Supplier, ditto (+)(P)

Agent, ditto (+) (C)

Customer, ditto (+) (C)

Average staff seniority (+)(HR)

Percentage of re-located staff (+)(HR)

Percentage of staff with disabilities (+)(HR)

Percentage of working hours spent on corporate volunteering (+)(HR)

10) Power

Turnover (+)(G)

Headcount (+)(620,3,600)(1)(HR)

Company share capital (+)(G)

Number of subsidiaries (+)(G)

Number of investee companies (+)(G)

Percentage of board members holding positions in other companies (+)(G)

Region-wide market share (+) (C)

Nation-wide market share (+) (C)

World-wide market share (+) (C)

Public opinion of power wielded by the company (+)(7,5,7)(1)(G)

Inputs (X)

1. Total expenditure, General Management (-)(150,5,145)(G)
2. Total expenditure, Commercial Department (-)(80,2,85) (C)
3. Total expenditure, Production Department (-)(70,-2,80)(P)
4. Total expenditure, Administration (-)(55,1,56)(A)
5. Total expenditure, Human Resources Department (-)(62,3,64)(HR)

Note that the Input indicators bear a minus sign (the less spent, the better) and they have no weighting factors inasmuch as they all refer to a respective monetary unit.

The above list is just one of many samples that could be adopted. But accepting a list of indicators may be the first and foremost action taken by any company management, for it involves setting the objectives to be met in a given period or, more graphically, programming the lights on the scoreboard to track and monitor management information.

Each company's list of indicators varies, naturally, depending on size, nature, problems and expectations for the future. The basic principle to bear in mind is that such lists should be meaningful, useful, simple and comprehensible for the largest possible number of people inside and outside the company. The number of indicators should be minimised without excluding relevant information (the more meaningful the better).

Another problem to be solved in advance is the comparative importance of the different indicators. This can be done by assigning a relative weight to each (on a scale of 1 to 10, for instance). The result would be that a company's overall balance sheet would no longer be a simple sum such as a+b+c ... +n, but rather that same sum, duly weighted with the respective factors or "p": ap1+bp2+cp3+... n+pn. These items, moreover, would be summable because the indicators would have been previously standardised to an indexed form.

It goes without saying that both operations, indicator selection and weight assignment, should be performed with a maximum of participation from the company's management staff, employee representatives and even external consultants with expertise in the area, who might be commissioned to submit an initial proposal for consideration. Delphi methodologies for inter-subjective expert agreement yield excellent results by combining different participants' points of view and even providing mathematically sound solutions.

3. The new Corporate Balanced Scorecard

The following table is a simplified Corporate Balanced Scorecard (CBS) developed around no more than the 20 indicators listed above, which show, in the central parentheses, the sums that reflect management area results. As explained above, the first figure represents the previous period's achievement, the second the improvement in percentage points and the third achievement in the second period. The quantitative information provided by this Balanced Scorecard will also be used below to measure business efficiency.

Table 1: A Simplified Corporate Balanced Scoreboard (CBS)

Indicator	Sign	P.1	F.I., %	p.2	A.I., %	Dev	Wt.	WTED .DEV.	Iml.	Dept
(1)	(2)	(3)	(4)	(5)	(6)	(7)	(8)	(9)	(10)	(11)
Output										
Absentee-ism, days	-O	215	5	217	-0.9	- 5.09	5	-29.5	0.99	H.R.
Opinion ab. Health	+S	8	5	11	37.5	32.5	1	32.5	1.37	H.R.
Productivity	+O	120	6	125	4.2	-1.8	9	-16.5	1.04	G
Opinion ab. Income	+S	7.7	5	10	29.9	24.9	1	24.9	1.3	H.R.
Alloc. for reserves	+O	20	5	21	5	0	4	0	1.05	G
Opinion ab. Security	+S	6	5	5	- 16.7	- 21.7	10	- 216.7	0.83	G
Informative meetings	+O	12	0	15	25	25	1	25	1.25	H.R.
Opinion ab. Freedom	+S	7	5	7	0	-5	6	-30	1	H.R.
Staff w/ univ. degree	+O	25	10	35	40	30	2	60	1.4	H.R.
Opinion ab. Knowledge	+S	8	10	8	0	-10	8	-80	1	G
Profit sharing	+O	37	5	38	2.7	-2.3	7	-16.1	1.03	H.R.
Opinion ab. Equity	+S	6	10	6	0	-10	6	-60	1	H.R.
Waste generated	-O	15	5	10	33.3	28.3	2	56.6	1.5	P
Opinion ab. Pollut.	+S	7	5	7	0	-5	2	-10	1	P
Employee suggestions	+O	300	5	320	6.7	1.7	7	11.7	1.07	H.R.
Social wrkplce climate	+S	6	10	6	0	-10	8	-80	1	H.R.
Customer complaints	-O	38	5	30	21.1	16.1	1	16.1	1.27	C
Opinion ab. Prestige	+S	6	4	5	- 16.7	- 20.7	10	-207	0.83	H.R.
Headcount	+O	620	3	600	-3.2	-6.2	1	-6.2	0.97	H.R.
Opinion ab. Power	+S	7	5	7	0	-5	1	-5	1	G
Total			113		167. 8	54.8	92	- 529.9	21.9 0	

Average			5.7		8.4	2.7	4.6	-26.5	1.10	
Input										
G.1		150	5	145	3.33	- 1.67			1.03	G
G.2	-	80	2	85	- 6.25	- 8.25			0.94	C
G.3	-	70	-2	80	- 14.2	- 12.2			0.87	P
G.4	-	55	1	56	-1.8	-2.8			0.98	A
G.5	-	62	3	64	- 3.22	- 6.22			0.96	H.R.
Total		417	9	430	- 22.1	- 31.1			4.78	
Average		83.4	1.8	86	-4.4	-6.2			0.96	

In this table, the numbers of the columns denote the following:

(2): Signs "+" or "–" according to the desirability of the indicator, and "O" stands for "objective" and "S" for "subjective".

(3): Forecast output in absolute figures.

(4): Forecast percentage of improvement

(5): Actual output in absolute figures

(6): Actual percentage of improvement

(7): Percentage of deviation produced

(8): Weight or relative importance of indicators

(9): Weighted deviation

(10): Improvement index=(3)/(5) or (5)/(3) according to the desirability of indicators

(11): Department responsible

The overall results of this CBS, based on the hypothetical data, were as follows:

(1) The forecast improvement in percentage for all the company *outputs* was 5.7 (column 4).

(2) The actual improvement, however, was 8.4 (column 6), for a favourable overall deviation of 2.7% (column 7).

(3) The forecast improvement in percentage for all the company *inputs* was 1.8

(savings were planned) (column 4 under Inputs).

(4) Actually, however, improvement in this regard was negative (the -4.4% in column 6 under Inputs): in other words, the company spent more than planned: 430 million instead of the 417 budgeted, for a *negative deviation* (-6.2).

(5) The total percentages (columns 4 and 5) can be used to calculate a first estimate of company efficacy by comparing the following ratios:

(a) Forecast efficacy (Tf)=(100+5.7)/(100-1.8)=105.7/98.2=1.08

(b) Actual efficacy (Ta)=(100+8.4)/(100+4.4)=108.4/104.4=1.04.

In other words, company forecasts called for positive "transformation" at a rate of 8% (1.08) but the rate actually attained was only 4% (1.04).

(6) After the results are weighted with the figures in column 8, however, the result is negative (-26.5), meaning that positive deviations were attained in indicators with scant relative weight, while the results were negative for important indicators with weights of 8, 9 or 10. The deduction is that management neglected matters of greater relevance to attend to questions of minor importance.

(7) Column 10, which relates columns 3 and 5, again shows, with a total index of 1.10, that unweighted results were 10% higher, not with regard to the forecast, but with regard to the previous period. This column gives the degree of dynamic (=over time) improvement or regression.

(8) Under Inputs, the same column shows that the result of having incurred greater expense than planned was a negative deviation of 96-100=-4%.

These results can be subdivided by department or area of responsibility, although this is not shown in the above illustration, which, as noted, is limited to just 20 indicators. Results can also be expressed graphically to facilitate evaluation.

4. Calculating business efficiency: when is a company efficient?

The foregoing discussion has been leading up to the key concept of this paper, "business efficiency". The author has consistently sustained that companies cannot provide social welfare if they are not efficient. Hence, the definition of "business efficiency" is the final point of the present paper, the point where it will be sustained that the social welfare (SW) provided by a country's companies depends largely on the overall business efficiency (BE) attained.

Indeed, it would be to little avail for companies to engage in "philanthropy" or reach suitable levels of so-called "corporate social responsibility" if by so doing they were to jeopardize their overall efficiency. This concept is regarded here to be more inclusive, from a theoretical perspective, because it is based on the Values/Stakeholder matrix and systematic inter-company comparison or benchmarking, both of cardinal importance to company durability.

Moreover, the enormous economic effort that companies (particularly Spanish companies) need to devote to research and development may prove to be incompatible with much of what may be regarded as "social" expenses or related items. Under such circumstances of technological disadvantage, which must be remedied to ensure overall economic development, it may be correctly sustained that companies' true contribution to a society's welfare consists in prioritising – absolutely and above and beyond many other aspects and expenses of apparent social significance – the implementation of relevant scientific research and technological innovation programmes. Consequently, indicators such as those presented in the general list under value 5, "Knowledge" in the above Corporate Balanced Scorecard (CBS), should carry a much higher relative weight than others, either in number or relative individual weight.

But not only do the indicators focus on the Knowledge value: the entire CBS has been designed around the need for progress in innovation and new technologies, and for knowledge to flow to and from staff, suppliers, customers and company networks in an effort to join forces for their mutual benefit through their responses, ideas, suggestions and projects.

The concept of "efficiency" presented in this paper should, therefore, yield a series of theoretical dimensions of overall importance, the operational definition of which should be highly demanding in terms of competence-related content. What this means is that companies labelled as "efficient" under this approach (i.e., those with an overall index greater than one, as discussed below) will be able to ascertain that they in fact are presently and will continue to be efficient in the future, and that such efficiency is measured with respect to relevant competitors. In the end, then, given the predominant role of private enterprise in modern society, the concept of Social Welfare will hinge almost exclusively on Business Efficiency, the backbone of any national economy. Given these assumptions, the relevant question is: when can a company be said to be efficient?

Providing an appropriate or scientifically valid answer to this question involves solving certain basic theoretical problems. The first necessitates practically denying a more or less accepted hypothesis according to which the concept cannot be measured. Such a hypothesis, as formulated below, can be found throughout scientific literature on organisations: "Efficiency in organisations cannot be measured or calibrated for want of a general comparative model."This hypothesis has been supported by most scholars addressing the subject (Edwards et al., 1986).

Nonetheless, many authors have attempted to measure organisational efficiency empirically. Miles (1980), for instance, used 29 measurements; Campbell (1977) 30 criteria; Mahoney (1977) 114 variables; and Seashore & Yutchman (1967) 76 different indicators. Some authors (Dalton and Kesner, 1985) even claim that the number of possible measurements is nearly infinite, while all stress the difficulty involved in standardising measures for comparison. Generally speaking, positions range from those (such as Goodman, Atkin and Schoormann, 1983) who propose a moratorium in the analysis of organisational efficiency until better inter-subjective conditions are in place, to those who propose definitively abandoning the idea in view of the utter impossibility of ever reaching an agreement (such as Hannan & Freemann, 1977).

There are, naturally, authors (such as Morgan, 1980) who believe that such an agreement is not impossible or who argue that the decisive importance of the concept precludes abandonment if the aim pursued is to understand and improve business organisations (such as Peters & Waterman, 1982; Handy 1993 and the Total Quality Control movement in general. More recent but likewise theoretically disoriented approaches can be found in Mullins, 1996).

In short, from the earliest attempts quoted above to the most recent papers of which this author is aware, such as Puig-Junoy (2000), Surruca (2003) or Vergés (2004), which have brought important advances in the definition of the concept, the hypothetical impossibility of the endeavour may still be said to be accepted. The explicit rationale for this hypothesis is based on the lack of a general comparative model able to generate the necessary agreement among experts.

The present paper has, however, attempted to show that such a model does exist, subject only to deployment of the respective effort to attain theoretical integration, thereby eroding the scientific basis for the above hypothesis (see the Reference Pattern of Values).

Assuming, for the time being, the existence of a valid model, the second problem consists in the broad polysemy of the term "Efficiency" itself, and the equally broad overlap with other synonyms, discussed by several authors. In this regard, confusion may be said to abound among terms such as "Efficacy" (when the company reaches its stated objectives); "Effectiveness" (when its stakeholders accept company results); "Efficiency" (when the company maximizes overall rationality); "Profitability" (when the measure is the capital gains generated); "Productivity" (when production is related to the number of employees); "Success" (when the ultimate long-term ends pursued are attained); "Growth" (when turnover increases from period to period); "Development" (when certain desirable levels are reached); "Excellence" (when financial profitability and company enlargement prevail) and so forth. Clearly, semantic chaos reigns around the term Efficiency. Webster's Third International Dictionary is of scant assistance: it defines *effective* as that which is "able to accomplish a purpose" and *efficiency* as "suitability for a task or purpose", i.e., essentially the same.

It may be deduced from the ideas associated with all these terms that none of them would, by itself, fully cover the idea of business's contribution to *social welfare*, including therein not only the idea of the attainment of the initial aims ("efficacy" for many), but also their achievement at the lowest possible cost or effort (what others call "efficiency"), with year-on-year improvements ("advancement" or "development") or even with employee and stakeholder conformity with or acceptance of what is obtained (usually termed "Effectiveness" in political science literature). What concept could encompass at least these four component principles of desirable business behaviour? The English and French term "performance", for instance, would appear to represent a more global vision of the results of business endeavour, but there is no true equivalent in other languages, such as Spanish.

Given the obvious need for a concept that would express the ideas of Efficiency, Efficacy, Results and so on, from as global a perspective as possible (integrating economic and social aspects in the broadest possible sense, in keeping with the term "social welfare"), perhaps the most viable solution consists in redefining some of these concepts. *Efficiency*, for instance, would not be reduced to the almost dangerously narrow "ends reached/means deployed" ratio, but expanded to include most of the meaning of the other terms described above. If the above question is to be answered with any rigour, company complexities must be taken into account in all their dimensions, an undertaking which, in principle, is simply a matter of adding versus subtracting perspectives until an integrated presentation of the concept is attained.

The first operation would consist in re-labelling the "ends reached/means deployed" ratio, which has been termed "efficiency", to release this term from the confines of its current content. What might this ratio, which signifies achievement of a result pursued at low cost, effort, or in other words low *energy consumption*, be called? The term *ecological* might prove to be suitable, inasmuch as the intention is to minimise such consumption. Albeit provisionally, the adjective "ecological" might be adopted to designate business behaviour exhibiting a suitable Input/Output ratio. This, in short, would represent the classic Output/Input ratio as the essential result of any company's or even any social system's transforming action.

Once the concept of Efficiency is freed from the narrowness of the above ratio, a more complex, operational and at the same time quantitative definition of Business Efficiency (BE) might be advanced. Such an endeavour must be preceded by a brief introduction to an eminently sociological approach to the business system, again in keeping with the "social welfare" concept; therefore, we should dispose of a list of the different stakeholders comprising the complex world of business relationships. The Reference Pattern of Values, in turn, is given in Table 1 (see chapter 3) and the resulting Matrix (Values -rows- against stakeholders –columns-) would relate stakeholder expectations to the action the company must undertake to meet them. This provides an overview of the powerful interests that shape business activity and its functional dependence on such stakeholders, all of whom expect to obtain something from the company: salaries, dividends, products, services or taxes, at times as keenly as if such items were as essential as the very air they breathe. Like it or not, the central role of private enterprise as the mainstay of modern society cannot be denied; nor can the dependence of social structures as a whole on private profit be ignored, even if viewed from more critical and countercultural perspectives.

5. The six requirements of the concept of "Organisational Efficiency"

Taking these conceptual bases as a point of departure, the initial question would have to be re-formulated in a more general and concrete manner: *When is a company efficient?* Initially, as argued above, when it is simultaneously "Ecological", "Efficacious", "Effective" and "Incremental". And it must be all these things with respect to relevant competitors, for nothing can be said to be good/bad, tall/short, ugly/beautiful and so on unless in comparison to some reference. A company may be highly ecological, efficacious, effective and incremental, yet still be the least ecological, efficacious, effective and incremental of all companies in the same industry and of comparable size. The definition of the new concept calls, then, for the introduction of at least one more dimension: *internal/external* that compares company results to those of its (relevant) competitors. Although this information cannot always be readily gathered, it is becoming increasingly more accessible on the Internet where many or most of the data needed for such assessments can be obtained.

The conceptual model for Business Efficiency (BE) would, therefore, be defined by the following six propositions:

1) A company is efficient if and only if it is *ecological* (attains a desirable Input/Output ratio). 2) A company is efficient if and only if it is *efficacious* (obtains what it plans to obtain). 3) A company is efficient if and only if it is *effective* (its results are accepted by its stakeholders). 4) A company is efficient if and only if it is *incremental* (its results are an improvement over the preceding period, i.e., the positive factors grow and the negative factors decline). 5) A company is efficient if and only if it is *profitable* (earns suitable financial profits). 6) A company is efficient if and only if it is *adapted* (it is at least as ecological in its basic ratio between "Outputs" and "Inputs" as its relevant competitors, on average).

All of the foregoing is based on the assumption that the set of indicators used validly operationalises the theoretical Reference Pattern of Values and Company Stakeholder models. Otherwise, the utility of the approach would have to be challenged or the approach redefined.

The Business Efficiency Index would, then, be formulated from the following indices:

1. *Ecological Dimension* (T)

This is the ratio between *outputs* (Y) and *inputs* (X).

Therefore, T=Y/X,

where "Y" is the average of the percentage improvements obtained in the "Output" indicators and "X" the percentage improvement in the "Input" indicator, both with respect to the preceding period.

2. *"Efficacy" dimension* (E)

This is the ratio between FORECASTS and ACHIEVEMENTS

Therefore,

E=Ta.α/Tf

where "Ta" is the actual, "Tf" the forecast ecological dimension and "α" a coefficient of correlation.

3. *"Effeciveness" dimension* (Ef)

This is the ratio between SUBJECTIVE and OBJECTIVE

Therefore,

Ef=Y(S)/Y(O)

where Y(S) are the outputs as perceived by stakeholders and

Y(O) the Outputs actually attained.

4. *Incremental dimension* (I)

This is the ratio between PRESENT and PAST

I=(I1+I2+ … +In/n)

where I1, I2 and so on are "Output" indicators.

5. *"Adaptation" dimension* (A)

This is the ratio between the COMPANY and its COMPETITORS

Therefore,

A=Ta/Te

where Ta is the company's ecological dimension and Te the ecological dimension

corresponding to its relevant competitors.

6. *"Profitability" dimension* (P)

This is the ratio between the company's gross profits (CGP) and the average gross profit (eGP) earned by its relevant competitors.

Therefore,

P=CGP/eGP

where CGP is the company's gross profit and eGP is the mean

The organisational efficiency index (OEI) would be:

OEI=(T+E+Ef+I+A)/5

While the Business Efficiency Index (BEI) would constitute the integration of the two indices:

BEI=(EO+P)/2

The averages calculated are arithmetically correct inasmuch as the values found for all the resulting expressions hover around "1". In all cases, results > 1 indicate "high efficiency", whereas results < 1 mean "low efficiency", which may be interpreted for each dimension examined in terms of the respective deviation from "1".

The data in the sample Corporate Balanced Scoreboard above can be substituted into these equations to clarify the content of the concept and illustrate its simplicity.

The only difficulty in calculating these indices in the real world lies in the availability of the quantitative information required, which may not be immediate, particularly as regards the data on competitors. But this is one of the challenges that the information society poses to modern companies.

6. A practical example using sample corporate balanced scoreboard data

"Business Efficiency" can be calculated by substituting the data given in Table 1, plus the information gathered via surveys and facts on the competition, into the above expression. The following supplementary information (also hypothetical) is provided:

a) Company profitability (gross profit over turnover) = 16%.

b) Competitor profitability (average gross profit over turnover) = 15%.

c) Average value of "T" for competitors = 0.85.

Calculating the various dimensions:

Actual ecological index (Ta)=Y/X=(100+8.4)/(100+4.4)=108.4/104.4=1.04, where: "Y" stands for Outputs, "X" for Inputs and 8.4 and 4.4 for the respective average percentage improvements. These results mean that the company obtained a value of 104 from inputs worth 100, the transformation of which yielded a profitability of 4% in terms of energy.

Efficacy index (E)=Ta/Tf=1.04/1.08=0.96,
where: "Tf" is the forecast ecological index (100-5.7)/(100-1.8)=105.7/98.2=1.08. This result relates forecasts to actual achievements, for a 4% loss, although here the reference is the budget.

Effectiveness index (Ef)=ImI(S)/ImI(O)=10.33/11.89=0.89,
where: ImI(S) is the sum of the Improvement Indices (column 10) for the subjective indicators (S) and ImI(O) the sum of the Improvement Indices (column 10) for the objective indicators (O).

Be it said here that ImI(O) represents the true Outputs based on statistical facts, whereas ImI(S) represents stakeholder opinion of the level reached. The results show that the perception of overall improvement is 10.33/10=1.03, whereas the actual overall improvement was substantially higher, 11.59/10=1.16. This is an indication of company ineffectuality in explaining its actual results to its stakeholders. Since an Effectiveness Index > 1 would signify "manipulation" and < 1 "lack of information", the obvious lack of information attributable to the company in this case reveals a need to revise its communication procedures.

Increment Index (I)=(I1+I2+ ... +In)/n=(0.99+1.37+.......+1)/20=1.10
Here the succession of indices is taken from column 10 and the result means that achievement in the most recent period in terms of the 20 indicators taken as a whole was 10% better than in the preceding period. Inasmuch as the improvement indices (ImI) are obtained from the ratio between P1 and P2 (columns 3 and 5), this result denotes net positive "growth" in company

activities overall.

Adaptation Index (A)=Ta/T =1.04/0.85=1.22

Here 0.85 is the value of Ta for all the relevant competitors, obtained by benchmarking.

This outcome means that the results obtained by the company were 22% better, as measured by the 20 CBS indicators as a whole, than the figures recorded for its competitors.

Profitability Index (P) = CGP/GP = 16/15=1.07

Here 16 and 15 are the gross profits obtained from company books and through benchmarking, respectively.

This result means that the company was 7% more profitable than its competitors.

Consequently, the Organisational and Business Efficiency Indices (OEI and BEI, respectively) are:

OEI = (T+E+Ef+I+A)/5=(1.04+0.96+0,89+1.10+1,22)/5=1.04

What this means is that, viewed as a pure, non-financial organisation, the company was 4% more efficient than other organisations engaging in the same business.

Finally, the total Business Efficiency Index (BEI) comes to:

BEI = (OEI+P)/2=(1.04+1.07)/2=1.05

This overall result is interpreted to mean that the company performed 5% better than the industry average in all the economic and social aspects considered, taken as a whole. Once again, the foregoing figures are all hypothetical and serve no other purpose than to provide material for the present practical exercise.

Conclusion

From these calculations, it seems clear that is perfectly possible to work out the concept of *organisational performance* in a comparable way, both in time within the organisation, and in the space of other comparable organisations. Summing up, the main characteristics of this concept are: a) It is theoretically founded since it uses the Referential Pattern of Values in order to cover basic universal human needs; b) It therefore integrates economic and social aspects; c) It integrates objective statistical information (facts), and subjective feelings of people (opinions) about these facts; d) It compares the forecast performances and the achieved performances; and e) It uses standardised data within the interval 0-100, so that axiological comparable profiles can be drawn or sketched. Moreover, the six dimensions of the concept of Organisational Efficiency suggested here (Ecological, Efficacity, Effectivity; Incremental, Adaptation and Profitability) as theoretical requirements would oblige organisations and corporations to include in their Balance Score Boards not only social and ecological aspects but also a presentation of their results in relation to comparable organisations.

References

Campbell K.S. (1977) "On the Nature of Organisational Effectiveness". In *New Perspectives on Organisational Effectiveness.* ed. by P.S. Goodman P.S. & Pennings J.M. San Francisco CA: Jossy Bass.

Dalton R.D. & Kesner I.F. (1985) "Organisational Performance as an Antecedent of inside/outside Chief Executive Succession: An Empirical Assessment". In *Academy of Management.* Journal 28 (4).

Goodman P.S., Atkín R.S. & Schoorman F.D. (1993) "On the Demise of Organisational Effectiveness Studies". In *"Organisational Effectiveness: a comparison of multiple models".* ed. by Cameron K.S. & Whetten D.A. New York: Academic Press, 163 - 183.

Handy C.B. (1993) *Understanding Organisations.* London: Penguin.

Hannan M.T. & Freemann J.H. (1997) "The Population Ecology of Organisations". *American Journal of Sociology* 82, 924-64.

Horvath & Partners, (2003) *Dominar el cuadro de mando integral: Manual Práctico Basado en Más de 100 Experiencias. Barcelona: Gestión.*

Kaplan R.S. & Norton D.P. (1997) *Cuadro de Mando Integral (The Balance Scorecard).*

Mahoney T.A. (1977) "Managerial Perceptions of Organisational Effectiveness". *Administrative Science Quarterly* 14.

Morgan G. (1980) "Paradigms, metaphors and puzzle solving in organisational theory". *Administrative Science Quarterly* 25.

Morgan D. & Zeffane R. (2003) "Employee involvement, organisational change and trust in management". In *International J. of Human Resource Management.* ed. by Taylor & Francis.

Mullins J. (1996) "*Management and Organisational Behaviour*". Pitman Publishing.

Peters T. & Watermann R. (1982) *In search of excellence.* Warner Books.

Puig-Junoy y Dalmau E. (2000) "¿Qué sabemos acerca de la eficiencia de las organizaciones sanitarias en España?: una revisión de la literatura económica". XX Jornadas de Economía y Salud Palma de Mallorca. 3-5 mayo.

Seashore S.E. & Yuchtman, E. (1967) "Factorial Analysis of organisational performance". *Administrative Science. Quarterly* 12.

Surruca, J. (2003) "Gobierno de la empresa y eficiencia en organizaciones orientadas a los interesados: una aplicación a las cajas de ahorro y a las cooperativas de Mondragón", Tesis doctoral, U.A. de Barcelona.

Vergés J. (2004) "La eficiencia (productiva) y la relación "principal/agente" en el caso de las Eps". Documento de trabajo.

4

Efficiency and Non-Profit Organisations

By Chaime Marcuello-Servós

Abstract

This paper explores a theoretical framework for discussing the concept of efficiency in the ambit of Non Profit Organisations (NPOs) and puts forward a model of Social Efficiency Indicators (SEIs). Its origin is in resolving a practical problem: *how should we distribute our available resources between the NPOs?* The unanimous answer is to look for the best social performance. The rhetoric of efficiency is used as an implicit normative axiom: *"you must choose the most efficient social behaviours".* This is a consequence of a simple and direct application of economic calculations to their idea of social efficiency. But we cannot strictly apply this algorithm. According to our empirical research, a SEIs model is proposed as a guide that goes beyond the assumption that the economy is frictionless and economic successes socially perfect. A final theorem about what social efficiency is has not been found. A model is sought but it is only asymptotically reachable.

1. Introduction

The concept of Non-Profit Organisations (NPOs) includes a heterogeneous and wide universe of entities, activities and groups of people. It is also referred to as the Third Sector[1], the *Anglo-Saxon* approach, or in a similar sense as Social

[1] For further information on this point see: http://www.istr.org/, where it states: *"The International Society for Third-Sector Research (ISTR) is a major international association promoting research and education in the fields of philanthropy, civil society and the nonprofit sector. ISTR reflects the growing worldwide interest in Third Sector research and provides a permanent forum for*

Economy[2], *the Continental* approach, (Bellostas, 2001), (Marcuello, 2007), (Salamon & Anheier, 1994), (Weisbord, 1975) and it is accepted that it is a collection of distinct organisations with a common focus: non-profit orientation. Basically, this means there is no distribution of benefits to the "owners" and/or stakeholders, along with other aspects such as voluntary membership, independent government, and so on. Analysis of this sector, together with discussions and debates, is increasing constantly, with a large amount of papers and books being produced. However, the aim of this paper is not to comment on these but to focus on efficiency in NPOs in theory and practice.

I would like to explore a theoretical framework for discussing the concept of efficiency in the ambit of Non Profit Organisations (NPOs). It arises from an empirical research process that started more than ten years ago (Marcuello, 1996). This aim has its origin in a practical problem: *how should we distribute our available resources and budgets between NPOs?* This is a recurring question, especially from policy makers, large philanthropic donors and ordinary citizens, and also in academic circles. Usually, the unanimous answer is to look for the best social performance. They choose the best option and whichever of these particular "best options" is legitimated according to their perception of social efficiency outputs is chosen. The rhetoric of efficiency is used as the main argument, i.e. as an implicit normative axiom: *"your choice must be efficient social behaviours".*

This is a consequence of a simple and direct application of economic calculations to their idea of social efficiency. However, one cannot strictly apply this algorithm, because what social efficiency means in this context could not be arrived at solely in terms of money or traditional economic analysis. This[3] empirical research within Third Sector organisations shows that what is needed is a change of perspective. This implies a distinction between an *organisational level* and an *evaluation or accountability level.* The former refers to internal

---

international research, while at the same time building a global scholarly community in this field"

[2] For further information see: http://ec.europa.eu/enterprise/entrepreneurship/coop/index.htm

[3] The author is a member of the Third Sector's Social and Economics Studies Group, GESES: http://geses.unizar.es.

structures, human resources management, financial matters and any other relevant item where it would be possible to calculate costs and relations between inputs and outputs. The latter is the assessment of effects and results. Both dimensions overlap many times and have evolved with regard to accountability and what Power (1997) called the "audit society". Thus, the evaluation level is the ambit where the Social Efficiency concept properly gains a relevant position. It could be understood as a result of the social impact of action and the best legitimation for taking decisions.

This is the context of our Social Efficiency Indicators (SEI) model. Our SEI model is put forward as a guide to describe social efficiency of NPOs, assuming that the economy is frictionless and economic successes and their outputs socially perfect. Our SEI model has been tested in order to develop *social audits* in the Spanish Third Sector. In this paper, I will present the theoretical discussion background, the process and the SEI model itself. Bearing in mind that a final theorem about what social efficiency is has not been found, a complete model is sought but is only asymptotically reachable.

2. Efficiency and its fuzzy aspects

There is a commonly accepted meaning of efficiency. According to Tang (1997: 461-462); *"efficiency refers to the economically appropriate allocation of resources. It is the relationship between inputs and outputs allowing for influence of factors outside the control of the agencies at issue. The most efficient arrangement is the one that produces the greatest output per unit of input, for example, the lowest cost for a given level and quality of service. It is easier to obtain a precise measure of the cost of inputs than of the quality and effectiveness of outputs. Thus, there is a built-in tendency for measures of efficiency to be better at comparing costs of input than the equivalence of output"*. This way of understanding efficiency shows a complete similitude with *engineering thinking*: ratio of energy and time, results and effort or output and input. It seems that if we apply this to social systems, we only need to distinguish processes, elements, operations, targets and scores. The efficiency of the system is simply calculated using the same mechanism, the same ratio, and the same logic; and it is probable that we can respond using the numbers and figures of these ratios. It is a *quantitative approach* based on a relationship of items. It is a relationship between means and results.

However, difficulties arise when we want to describe and/or measure the social dimensions of "efficiency". Firstly, there is not a canonical definition or way to test it. Secondly, it is then necessary to decide which aspects and objects to describe and measure. And thirdly, the social dimension is greater than the economic. This is known as the "embeddedness" condition (Granovetter, 1985) and has to be considered from a *qualitative approach*.

For instance, if we think in terms of political campaigns, the larger the number of votes in the ballot box the more effective the campaign. Its quantitative efficiency in economic terms is calculated by looking at the lowest investment. But we know that this *effectiveness* is not equivalent to an *efficiency* pattern. And, in terms of the social system, in many instances the most numerous is not always the best option to maintain a prosperous future; even if it is the cheapest campaign. We need to apply a qualitative approach to the production of values in the system, as Parra-Luna (2000) (2002) suggests with his Axiological System Theory (AST). This is in keeping with Falkenberg, (1998: 6):*"concerns for efficiency must often yield to concerns of equity (justice and rights)".*

We should not let the organisational level overlap with the evaluation dimension. We need to make, at least, two distinctions: one, between efficiency and effectiveness, the other, between quantitative and qualitative. It is useful to consider that effectiveness is observable in the benefits of organisational outcomes. In addition we need to take a long-term perspective in order to speak about the NPO's social efficiency, if we are reconsidering evaluating its social consequences.

In this sense, authors such as Sandefur (1983: 449) consider that *"organisational efficiency may generally be defined as the value of benefits relative to the costs of acquiring those benefits".* However, the *evaluation or accountability level* needs to describe the social impacts inside the NPO in context. We need to listen to the views of its stakeholders. Social efficiency includes a requirement of social impact assessment and management accountability. We need an implicit statement of what is best in social dimensions.... But is this achievable?

Social efficiency is not simply solved as a Paretian optimum because the social dimensions of the NPOs are more complex than a problem of economy where it is possible to apply the Pareto model. This is one of the difficulties. The Third Sector is defined mainly by its non-profit behavior, but there is more than one side to the coin; social efficiency is not guaranteed in non-profit entities. One needs to check their effectiveness in advance.

The rhetoric of "social efficiency" appears to justify and legitimate decisions to give support or money and in choosing one NPO as opposed to another. Thus, *common sense*, affectivity, mood, rational choice, tradition, family, necessity, and chance are possible explanations for the decision to give resources to a NPO. However, the current main argument is a syllogistic consequence of two assumed propositions: i) The best social performance has to be supported; ii). The best social performance is the one that is the most socially efficient. Therefore, everyone should support the most socially efficient NPOs. However, we find a circular explanation when we look more deeply into the concept. As Roth posits (2000: 110): *"the genesis of the efficiency standard contemplates theoretical constructs that cannot meaningfully be defined".* The term "social efficiency" seems an invocation to find reasons to support decisions. Nevertheless it is used and it has a meaning.

The notion of social efficiency plays an instrumental role. Perhaps the underlying principle is utilitarian thinking, in which case *"the pursuit of efficiency will enhance productivity and material well-being, be it in terms of purchasing power, or efficient use of raw materials in a production process or the degree to which a small investment can produce a great return for the owners. Thus productivity —or how long must a person work (input) in order to secure a certain basket of goods (output)— is one of many relevant measures of quality of life. In this respect, efficiency is good. Efficient economies are also able to provide their citizens with improved education, health services, etc., which in turn are elements of social equity. Efficiency and equity are in this respect mutually supportive"* (Falkenberg, 1998: 6). It seems that social efficiency is not possible without effectiveness and economic efficiency... and this is a controversial and ideological terrain. *"Concerns for social equity, efficiency and freedom are at the heart of the political debate in many countries",* (Falkenberg, 1998: 2). Could there be a common pattern or reference?

3. A Social Efficiency Indicators Model

If there is not a canonical definition of social efficiency, but people use the notion, one can take into account discourses and analyse the *social shared meanings* and symbolic universes surrounding it. One has to investigate the "social reality" by pursuing the meaning, uses and sense. At the beginning of this research process, the author decided to collect a range of different views and listen to social participants from a wide collection of NPOs. He spent several years (Marcuello, 2002) fine-tuning our model and comparing the various positions after asking people: *What is social efficiency? How do people use the concept? What does it mean?*

The empirical findings show how the discourses of people from a diverse range of starting points come closer together until they reach *cost/benefit logic.* This is akin to the advertisement or slogan: *"the best for the least".* It is the expression of a generalized calculus of means and results together with an assessment of the impact.This may be referred to as the *quantitative approach.* However, as was described earlier, this is not sufficient. There is a convergent trend with the *qualitative approach.* Consequently, what does the adjective "social" add?

Any efficient NPO has to accomplish its own goals; but the "social" issue appears to be something more. It refers to an external dimension of these organisations. A consensus point is summed up by the phrase: *society as a self-value*. This result has been described in detail in other works (Marcuello, 2002), the first approximation of which is to consider that any NPO is efficient in social terms if it is *society-building*.

Could this be generalised with regard to any cultural situation in any country? A first impression would lead one to say no, it is a culturally dependent answer. Every society, every social system has its own particular *cosmos* and pattern of thought. Moreover, in the same city, each political party has its own way of thinking when formulating a style for developing society and reaching the targets.

One could easily fall into this flow of paralysed thinking, but Max-Neef[4] (1994) resolved this some time ago by distinguishing between *needs and "satisfactors"*. Basic *human needs* are the same everywhere (Sen, 1985; 1987) (Max-Neef, 1994), the ways of solving or satisfying the needs, the "satisfactors", are culturally dependent. Following this argument, it is possible to put forward a "human" framework for social efficiency: everyone wants the best for his/her society. Thus, the social efficiency of NPOs is shaped by their participation in society (social system), which in turn shapes them internally (Marcuello, 2002: 289).

Social efficiency has to go beyond efficiency at an organisational level. It is more than an internal matter. If it refers to external effects; social efficiency appears as a reference to society and the improvement of it. This is a dynamic process of feedback and exchange. Subsequently any NPO will be socially efficient if it creates benefits for society as a whole or, from a utilitarian standpoint, at least for the majority of its members. However the complexity of NPOs also shows that social efficiency has a direct relationship with social profitability, even when an NPO is working in areas where it does not create value in everyday economic terms or focuses on non-market concerns. For many people, it is sometimes easier to define what social efficiency is not - by

---

[4] The original paper was: MAX-NEEF,M.; ELIZALDE, A.; HOPENHAYA,M (1986): "*Desarrollo a Escala Humana, una opción para el futuro*" Ed. CEPAUR-Fund. Dag Hammarskjöld. Development Dialogue, nº especial 1986, Mozala Suecia. Re-printed in 1994.

looking at the social outputs of NPO as uncoordinated effort, beneficence, individual aid, and a waste of all human resources.

Therefore, social efficiency refers to a set of facts, results, and targets of a particular NPO within a social system, where any observer could give his/her judgment after collecting the data.

It would even be possible to assign a numerical value. However the final score would be more useful if based on consensus. Nevertheless, the first priority is to focus on the appropriate targets for observation. This is similar to "reasoning with regard to health": the first step is to give a description and an anamnesis in order to facilitate diagnosis. After a long period of data collection, it is possible to produce a reliable protocol with correlations, valuations and decisions. However, the definition of health is always a fuzzy issue, even where a general and consensual proposal is established, for instance, by the World Health Organisation. For example, cholesterol indices and health scales will evolve according to the data and the feedback from empirical data[5].

Thus, there is not a final, universally valid definition of social efficiency. There are approaches that come close and a proposal for items to be observed. One can apply the term social efficiency indicators (SEI) to this collection of pointers and descriptors of different aspects and behaviours of the NPO in relation to its social system. This is the result of a process of researching and analysing the discourses of the NPO stakeholders.

---

[5] Experts introduce improvements and refine the perhaps 'figures' or 'tables' would be better omit. They create a new consensus and believe that a high concentration of triglycerides in the blood indicates an elevated risk of a stroke. This kind of correlation does not equate to a final definition of health and it is not a "law" whereby we can find a set of consequences according to previous causes.

Table 1: Social Efficiency Indicators

Nª	Indicator	Questioning//description
1.	**Context**	It is necessary to contextualize the diversity of NPOs
1.1.	▪ Origins…	Who? When? Where was it created?
1.2.	▪ "Surroundings"	Where is it operating?
1.3.	▪ Trajectory	What are the milestones of the NPO? Successes? Failures?
1.4.	▪ Results	Available data in any field, in coherence with the NPO
2.	**Means**	Description of the resources and characteristics of the NPO
2.1.	▪ Human	People: who?
2.2.	▪ Technological & material	"Hardware" and tools: what?
2.3.	▪ Financial structure	Money and funds:
3	**Addressee**	Who is it addressed to? What is the scope and repercussion?
4	**Social Networks**	Does the NPO participate and/or create networks, federations?
5	**Participants**	How many people are involved with or related to the NPO?
6	**Internal participation**	How does it work? What kind of internal life does it have?
7	**Reach**	Degree of openness/closeness to others and its repercussions
8	**Ways of doing**	What are the procedures and means used?
9	**Communication**	Channels of exchange and distribution of information
9.1.	▪ External:	Addressed to Public Administrations, Financial Organisations, Institutions, Third Sector, Business Sector, Networks
9.2.	▪ Internal:	Members, managers, workers, volunteers, etc.
10	**Plurality**	How does it deal with dissension and pluralism?
11	**Permeability**	Is it permeable to social needs, demands or desires? How?

It is necessary to point out that this model of SEI has an applied intention and forethought: *we need to describe*. We use this collection of pointers and descriptors of different aspects as a way of reporting the social performance of the NPOs.

4. Applying the model

Any measurement is a description of an observable pattern according to a unit of observation. So, the first step of any action of measuring —measurement— is an attempt to give a description of the observable and the second step is to make a comparison according to a reference. Both, the reference and the process of comparing, are social conventions; where one can find meaning, however, one can only understand their meaning if one has the specialised knowledge or interest to transform them into relevant information. Usually, measurements are figures or numbers: for instance, shoes are sized according to a numerical scale, an earthquake is described using the Richter scale or a body's temperature by the Celsius scale. We know that if we attribute a "grade 7" or higher, we are speaking of a destructive earthquake and similarly 40ºC is a high fever. The same occurs with cholesterol, myopia, speed, power, distance, money, and so on. We convert things outside of our brain into facts (Kjellman, 2007), these facts into observables and, subsequently, into descriptions that we usually present in the form of numbers. Nevertheless, do we need to follow the same process in the case of SEI? Do we need numerical figures in order to proceed? Could we only use descriptions as collections of words[6] to apply as a collection of indicators? How can this be done and why should it be?

The answers depend on the intention and goals of the observer and their social context. Numbers are very useful in a multitude of situations. Assigning quantities to descriptions solves many problems, but it also creates new difficulties. According to Parra-Luna (2007), quantification is an obligation and a solution for increasing the solidity of Sociological practice[7] and it is an obvious requirement when you try to measure anything. However, since any

---

[6] Spanish speakers know that the verb "contar" —to count— it is used to "tell stories" (words) and "to calculate" (numbers).

[7] "I usually propose my students the following example: between propositions (a) "this is light" and (b) "this weighs 15.2 grams" there is a huge difference (determining in science) because the first is not falsifiable and second does. This is, in language not quantified anything goes, and therefore worthless, or worth very little" (Parra-Luna, 2007, 2).

measurement represents a codification according to a pattern, readers must be capable of decoding this quantity in order to understand the sense and meaning of the quantitative description. If I want to know how many members an NPO has —or the incomes or salaries of its professionals—, then it is better to express this with numbers. However, it is not always the best solution to describe something using numbers —"numerología", (Ibáñez, 1986) — it may be necessary to use words —"palabrería", (Ibáñez, 1986)—: we have to use each as required.

In this case, SEIs are a way of describing the NPO's behaviour in order to give a picture of its performance. Obviously, the SEI model has to be embodied in a concrete NPO. It is necessary to elaborate with regard to its organisational profile, i.e.:

1) Entity definition: name, mission, vision and values;
2) Dimensions: types and number of members, affiliates, stakeholders, customers/beneficiaries, types and number of employees, financial aspects such as incomes and expenditures;
3) Internal structure: location (countries, regions, cities...), departments/units, task division, functions, management organisation (hierarchies, roles), controlling systems (externalized or internalized).
4) Management style: processes of decision making, grade of centralization, information systems, participation mechanisms, users, volunteers or the presence and participation of others.

This first step allows us to continue with the preparation of the report according to the SEI model. Next it is necessary to seek answers to a list of relevant questions. However, *who should do it?* Who should be responsible for describing the different aspects of the model? The most appropriate method would appear to be to transfer this task to an external observer or consultant. This is the most common practice; however, it is also possible to do it internally. The choice of method will depend on the purpose of the report. For instance: is it for an internal monitoring process or is it in response to an external control requirement? Usually, *audits* are different from *inspections*. The former could be for internal use or also for external information, but the later are always a response to a control authority or institution. This SEI model can be useful in both cases because it provides a pattern for collecting information about the entity's performance, which afterwards could be used for a variety of different purposes. These must be clarified at beginning of the process. The final purpose affects the way in which stakeholders answer. Sometimes responses change depending on who is asking the questions and what their goals are. And clearly this aspect is related to discussions regarding the pertinence and utility of a *sociocybernetical approach* as a second order observation (Marcuello, 2006), which obviously underlies the SEI model. However, it is not possible to review this in depth here.

A second consideration is to evaluate the timing; i.e. when it should be carried out and how many times. This SEI model is not meant to be applied only for measuring project compliance and for the evaluation of effects. This means that it is not a picture for comparison with a scale, to find out whether the values studied increase or decrease over a period of time. With this model, we have attempted to obtain and give information on the performance of any NPO as a society-building agent. Simply put, we are clarifying the performance provided by the entity. SEIs are not a model for an evaluation of projects in relation to an initial situation and other final objectives and results or inputs and outputs. They are a holistic way of considering the efficiency of any organisation and facilitating in-depth descriptions of its performance.

According to Parra-Luna (2002), (2007) it is necessary to transform words and values into numbers. It could be possible to apply a simplified translation considering the time before a project/program and the time after. In order to achieve this, however, a project must first include a set of axiological levels. Each one could correspond to the equation: $E_{(NPO)} = \frac{YN}{C}$, where: E = efficiency; Y= Previewed Level Rate of Value Satisfaction; N= Previewed Beneficiaries; C=Cost.

Subsequently, by means of a questionnaire, the beneficiaries of the project/program provide the data to substitute into the equation: $E_{(NPO)}=\frac{YN}{C}$. Furthermore, time *t* could be included by adding this item to *YN*, as follows: $E_{(NPO)}=\frac{YNt}{C}$.

It could be complemented by combining the Delphi technique with the judgments of experts. According to Parra-Luna, this facilitates a different and more "objective" level of information. This could be represented with a second version of the first equation as follows: $E_{(NPO)}=\frac{YSN}{C}$, where:

E = efficiency; N= Previewed Beneficiaries; C=Cost. Y= Previewed Level Rate of Value Satisfaction, (range 0-100); $Y=\frac{(y_1+y_2+\ldots+y_{10})}{10}$ this is determined by a Delphi technique and each "y" corresponds to Table 2.

S = Previewed Level Rate of Security, (range 0-100); $S=\frac{(s_1+s_2+\ldots+s_n)}{n}$ open to the sector of the NPO and, at least, with the next aspects:

s_1= NPO Managers Confidence Level

s_2 = NPO Historical Accomplishment

s_3 = Expenditures Periodical Controls (25, 50, 75, 100).

Table 2: Reference Pattern of Values

Source: Parra-Luna, 2002

No	Need	Value Pursued	Symbol
1	Physical and mental well-being	Health (S)	Y_1
2	Material sufficiency	Material wealth (RM)	Y_2
3	Protection against contingencies	Security (Se)	Y_3
4	Freedom of movement and thought	Freedom (L)	Y_4
5	Understanding of and command over Nature	Knowledge (C)	Y_5
6	Equity	Distributive justice (JD)	Y_6
7	Harmony with Nature	Environmental Conserv. (CN)	Y_7
8	Self-fulfilment	Quality of Activities (CA)	Y_8
9	Social esteem	Prestige (Pr)	Y_9
10	Influence	Power (P)	Y_{10}

The collection of data will allow the comparison of different NPO outputs and findings. This would appear to be a better approach to answering the main question of this paper. However, this process of codification would not be useful if we were to think in terms of the old debate of the *hermeneutical circle* between *Verstehen//Erklärung* (Ricoeur, 1988). The social performance of any NPO appears to be equivalent to the social action of an individual: i.e. *someone does something*. This has its outcomes, in the same way that a text requires a reader with a minimum level of literacy.

This SEI proposal aims to offer an applied way of determining and showing the social performance of any NPO using the most clear and simple language. This is a proposal which prioritises *performance understanding* over *explaining performance* by means of codification in numerical terms. Of course, a statement such as "excellent communication" or "high temperature" is less precise than 8 points out of 10 in communication or a 40° fever reading. Here the SEI model is not one single correlation of short adjectives to actions. It is built on *extended descriptions* of any of the indicators with the final objective of giving an in depth and dense picture of social performance.

Conclusion

We have to return to the opening question: *how should we distribute our available resources and budgets between NPOs?* The notion of Social Efficiency and its rhetoric are used to argue and select the best social performance in the case of NPOs. There is no simple choice and no direct answer. This is especially true in the light of an ideological statement which says that the NPO is better than state policy initiatives, because it is "more socially efficient". And paradoxically, the best NPO performance should get the support of policymakers and donors.. However, this is not the case. Theory is one thing, practice is another. This is the context of our Social Efficiency Indicators (SEI) model.

This SEI model is put forward as a guide for preparing descriptions to determine the social efficiency of NPOs. This model also allows the development of *social audits* that use it as a pattern for research. It is possible to apply it as a tool for improving the organisational level, and also for evaluation in relation to the accountability level.

To summarise, I hope to have demonstrated that the task of conceptualising the notion of Social Efficiency in the Third Sector is useful, and also that an attempt at application requires a long period of tuning and of building consensus. It is difficult to get down to the nitty-gritty of determining the meaning of social efficiency, and even more complex to apply it if a final theorem of what social efficiency actually is cannot be not found. A comprehensive model is sought, but it is only asymptotically reachable.

Acknowledgements

I want to thank Brenda Reed for her helpful support in reviewing and checking the English grammar and other comments about the text.

References

Bellostas, A. et al. (2002) Mimbres de un país. Sociedad civil y sector no lucrativo en Aragón.

Butler, R. (1983) "A transactional approach to organizing efficiency: perspectives from markets, hierarchies and collectives". Administration & Society 15 (3), 323-362.

Cheung, A.B.L (1996) "Efficiency As the Rhetoric: Public-Sector Reform in Hong Kong Explained". International Review of Administrative Sciences 62, 31-47.

Eecke, W. VER (1999) "Public goods: An ideal concept". Journal of Socio-Economics 28, 139 –156.

Falkenberg A.W. (1998) "Quality of Life: Efficiency, Equity and Freedom in the United States and Scandinavia". Journal of Socio-Economics 27(1), 1-27.

Granovetter, M. (1985) "Economic Action and Social Structure: The Problem of Embeddedness". American Journal of Sociology 91, 481-510. [Reprinted in Sociology of Economic Life. ed.by Granovetter and Richard Swedberg. Boulder: Westview, 51-77].

Ibáñez, J. (1986) "Perspectivas de la investigación social: el diseño en la perspectiva estructural". In García Ferrando, M; Ibáñez, J.; Alvira, F., El análisis de la realidad social. Métodos y técnicas de investigación social.

Kjellman, A. (2007) "On Meaning and Architecture of Language". Paper prepared for the 7th International Conference on Sociocybernetics, June 18-23, 2007 Murcia Spain. Available from: http://sociocybernetics.unizar.es/congresos/MURCIA/papers/arne-kjellman.pdf

Marcuello-Servós, C. et al. (2007) Capital social y organizaciones no lucrativas en España. El caso de las ONG para el Desarrollo.

Marcuello-Servós, CH. (2006) Sociocibernética. Lineamientos de un paradigma. Inst. Fernando el Católico- CSIC. Zaragoza.

Marcuello-Servós, CH. (2002) "Non-profit Entities and Social Efficiency: a Sociocybernetic Approach to Social Efficiency and its Measurement". International Review of Sociology 12 (2), July 2002, 283-294.

Marcuello-Servós, CH. (1996) "Identidad y acción de las Organizaciones No Gubernamentales". Gestión Pública y Privada, 1, 103-123.

Max-Neef, M. (1994) Desarrollo a escala humana: Conceptos, aplicaciones y algunas reflexiones. Barcelona, Icaria Montevideo, Nordan-Comunidad.

Parra-Luna, F. (2007) "El "pecado social" de la Sociología: una reflexión crítica desde la axiología sistémica" (mimeographed).

Parra-Luna, F. (2002) "Axiological Systems Theory: Its Applications To Organizations". Paper presented at Brisbane, Australia, July 2002. ISA XV World Congress of Sociology (mimeographed).

Parra-Luna, F. (2000) The Performance of Social Systems: Perspectives and problems. New York: Springer Science+Business Media.

Power, M. (1999). The Audit Society: Rituals of Verification. Oxford UK: Oxford University Press.

Parra-Luna, F. (1997) "From risk society to Audit Society". Soziale Systeme. Zeitschrift für Soziologische Theorie. Jahrgang 3 (1997), Heft 1, 3-22.

Ravallion, M. (2003) "On Measuring Aggregate 'Social Efficiency'". World Bank Policy. Research Working Paper 3166.

Ricoeur, P. (1988) Hermenéutica y acción. De la hermenéutica del texto a la hermenéutica de la acción.

Roth, T. P. (2000) "Efficiency: An inappropriate guide to structural transformation". Journal of Socio-Economics 29, 109 –126.

Salamon, L. M. and Anheier, H.K. (1994). The emerging Sector: An Overview. Baltimore: The Johns Hopkins University.

Sandefur, G.D. (1983) "Efficiency in Social Service Organisations". Administration & Society 14 (4), February, 449-468.

Sen, Amarty K. (1987) The Standard of Living. In: The Standard of Living: The Tanner Lectures on Human Values. ed.by Sen, A.K., Muellbauer, J., Kanbur,R., Hart, K. and Williams, B. Cambridge: Cambridge University Press.

Sen, Amarty K. (1985) Commodities and capabilities. Oxford: Oxford University Press.

Sen, Amarty K. (1983) "Liberty and social choice". Journal of Philosophy. vol. LXXX, nº1, January 1983, 5-28.

Sen, Amarty K. (1983) "Development: Which Way Now?" The Economic Journal nº 372, vol 93. Cambridge University Press, 745-763.

Tang, Kwong-Leung (1997) "Efficiency of the private sector: a critical review of empirical evidence from public services". International Review of Administrative Sciences 63, 459-474.

Weisbrod, B. A. (1988) The Nonprofit Economy. Harvard University Press.

Weisbrod, B. A. (1986) "Toward a theory of the voluntary nonprofit sector in a Three-Sector Economy". In The Economics of Nonprofit Institutions. ed.by Rose-Ackerman S.

Weisbrod, B. A. (1975) "Toward a theory of the voluntary nonprofit sector in a Three-Sector Economy". Altruism, Morality and Economic Theory, 171-195.

5

Providing and Measuring Efficiency of Innovation: a Complex Issue Requiring Requisite Holism

By Dr. Matjaz Mulej, Peter Fatur, M.A., Dr. Jozica Knez-Riedl, Andrej Kokol, MBA, Dr. Vojko Potocan, Damijan Prosenak, MBA, Dr. Branko Skafar, Dr. Zdenka Zenko

Abstract

The efficiency of the official invention-innovation-diffusion processes (IIDP) is less than 5%. Measuring it is complex: very many soft and hard impacts affect it in synergies and interdependently, and few attributes are routine. Authors use the contemporary process approach rather the older investment, result, or innovation approaches. They discuss measures in terms of the parties that influence the IIDP and measures in terms of assessment or measuring of results of influencing measures/efforts enhancing innovation. The impact of managers and other preconditions with indirect influences matter as much as final achievements. It is not enough to measure final achievements or tangible inputs. The requisite holism requires more synergistic viewpoints. Otherwise, the real causes will be difficult to detect and correct, and innovations will be difficult to implement. Over-simplification may cause complex wrong decisions and actions.

The selected problem and the solution proposed here address measures promoting invention-innovation-diffusion processes (IIDP), and measures for assessing IIDP's outcomes. The contemporary approach is no longer based on (1) investment in, or (2) results of, or (3) innovation, but (4) on the IIDP as a process. The selected viewpoint is the requisite holism of measuring. In principle, all influential factors should be monitored and measured and their impact and consequences assessed, in order to facilitate improvement. But very often, in IIDP their influences work in synergy, rather than separately; these and similar facts make them difficult to measure, perhaps even to assess. This is true concerning the daily routine, and even more so concerning IIDP and their outcome – innovation. Authors/owners of suggestions must decide to try and invest in advance, and then they can measure outcomes, once users decide which attempt qualifies as an innovation and which one is a failure.

Once we talk about measures, the following sentences by Einstein should be considered (Thorpe, 2003): 'When mathematical laws speak of reality, they are unreliable. And when they are reliable, they do not speak of reality'. (p. 117). 'Not everything that can be counted counts. Not everything that counts can be counted'. (p. 213). If a reader finds our contribution too complex, he/she should consider Einstein's next sentence: 'Everything should be kept as simple as possible, but no simpler'. (p. 41). Oversimplification hides the problem, as expressed in another quote from Einstein: 'Realism is only an illusion, but a very persistent illusion.' (p. 123). We will try to come as close to realism as possible. IIDP differs from routine in this respect: reality changes all the time; previous data can hardly ever be re-used.

1. Definition of Innovation and its Preconditions

In this contribution, we do not measure the efficiency of companies by comparison with others, like Parra Luna (2007) does, but the efficiency of IIDP, resulting in innovation. On the basis of the OECD's Frascati Manual of 1971, the European Union published its definition of innovation (EU 1995, quoted in EU 2000, p. 4). It is briefly summarised here:

Innovation is every novelty found beneficial by its users and therefore yielding benefit to its owners/authors as well. Innovation is neither limited to technological topics nor to incremental improvement any more, like it used to be in older definitions. It results from IIDP.

A simplified reading of the official definition of innovation as an outcome of Schumpeter's definition is more detailed (Skafar, 2006, p. 31), but it leaves innovation of management/governance and business style aside by including production, process, market, inputs and organisational innovations.

In order to not be too superficial, we will add Mulej's definition of IIDP. It distinguishes five steps of IIDP: invention – suggestion – potential innovation – innovation – diffusion; and three criteria: content – consequences – duty to work on innovation; they are put in synergy by networking in a 3-dimensional space (Mulej, in Mulej et al., 1994, p. xiv). There are five elements of IIDP: (1) business program items, (2) technology, (3) organization, (4) managerial style, and (5) methods of leading, working and co-working. Recently 5 more contents were added: (6) business style, (7) governance and management process, (8) VCEN (values, culture, ethics and norms), (9) our habits, and (10) others' habits (Mulej, 2013). Consequences can be (1) radical or (2) incremental. Duty to innovate may (1) exist or (2) not exist. 20/40 types result from synergy of these three criteria. All 20/40 of them matter and deserve measuring of efficiency.

These 40 types of innovation make a case of the dialectical system i.e. a synergetic network of all crucial and only crucial viewpoints (Mulej, 1974; Mulej 2013) What must be considered as an impact for an idea to become an innovation as an outcome of IIDP, and therefore deserves to be measured as inputs, is summarized in the "dialectical system of preconditions for innovation" (X denotes interdependence: no precondition may be missing; they work in synergy):

Innovation = (invention/suggestion X entrepreneurial spirit/entrepreneurship X management X requisite holism X co-workers X VCEN[8] aimed at innovating, co-operation and interdependence X competitors X customers X suppliers X socio-economic conditions X natural environment X incidental events/good luck)

For innovations of any kind to surface from IIDP, the above preconditions are interdependent and all of them must be created, which requires the measuring of the efficiency of their management. This is why, according to historical and current practical experience, the innovation of management is the most crucial one (Andoljsek-Mesner, 1995; Barabba, 2004; Huston and Sakkab, 2006; IBM, 2006; McGregor, 2006; Mulej, ed., 1984; Mulej et al., 1987; Reich, 1984; Rosenberg, Birdzell, 1986). It must enable a continuous IIDP as efficiently as possible. Even in countries with very high standards of innovation, people are not happy with their results of IIDP (Business Week, 2006; Dewulf, 2006; EU, 2004; Florida, 2005; Kettula, 2005; Sato, Kumagai, Tsukuda, Numata, 2005; etc.) and suggest new approaches and methods to help with problem-solving. In some emerging market economies, innovation/IIDP is still considered a political topic rather than a scientific one (Gu et al, editors, 2006). According to Business Week (2005), only 4.5% of innovation projects succeed, even in the top companies worldwide. Before ideas or inventions become projects, most of them disappear rather than become suggestions, which are at least recorded. Efficiency in choosing the projects matters, too. It should be measured at the end of the project realization and during each phase (Haug, 2007).

This complexity does not mean that creative activities should be free of any supervision and related measurement; this would leave them in organizational disorder. It only means that creative work does not profit from over bearing supervision of consumption of every resource, or from rigid determination of time to be spent at work, or similar bureaucratic mechanisms

[8] VCEN stands for values – culture – ethics – norms in interdependence, reflecting the irrational, but very influential 'right brain' of human personality. (Potocan et al, 2007).

that are otherwise helpful in routine operations and projects. The essence of supervision of creative activities is the encouragement of IIDP and the removal of avoidable obstacles that might make this work more difficult and hinder innovation. Such supervision should not be envisaged as a short-term, but rather as a long-term measure, and, like IIDP management, requires above-average creativity. (Srica, 1999, 199, in Gustin, 2007, 74-75). To make measurement more objective, reviewers and professionals are invited to assess results per phases, as well as the overall results (Gustin, 2007, 75). Enabling continuous IIDP, rather than allocating blame, is the aim of measuring, and so is stopping unpromising projects in good time.

2. Enabling Continuous Innovation in an Organisation

In every organization, the framework model for managing continuous IIDP will usually be the same in general terms, while details will require specialized analysis and planning. Drafting of vision, mission, policy, strategy, tactics, operation processes and organization is followed by decisions on and in all of them to create a basis for running the operations. We must consider, in addition, that the implementation of a strategy is at least as complex as its creation; it is here that we confront established old habits (Feucht, 1995). Checking of results makes it possible to intervene when and where it is needed in all IIDP phases; it also provides part of the basis for the next drafting. (Mulej et al, 1992). All phases matter, if innovation is to be a continuous source of survival. Hence all of them must be efficient and require measurement or assessment. The efficiency of the earlier phases of the management process seems to be measured at the end of the entire process, which tends to hide the real causes of both success and failure (Belak, 2003).

3. Requisitely Holistic Measuring of IIDPs, their Preconditions and Outcomes

3.1 The Human Preconditions

Several recent experiences (Fujimoto, 2007; Huston, Sakkab, 2006; IBM, 2006; McGregor, 2006) have shown that organizations must work on IIDP in a top-down approach for co-ordination of the huge necessary funds and human resources, but with no centralization, which would take too much initiative away from the people who actually do the work. In addition, they should involve the largest possible number of co-workers for following two basic reasons:

1) Many co-workers have a multitude of ideas which could lead to innovations, but do not. These co-workers are often not asked to express their criticisms and ideas, which never become suggestions etc. due to an obsolete non-democratic management style.
2) Many ideas never become innovations because they do not receive support from people who would have to give up their established knowledge and habits if they accepted the new ideas. However, people do not oppose ideas, but fight for them if they feel that they are authors or co-authors of an idea. Managers must enable their co-workers to be co-authors.

Thus, conclusions can be drawn as follows:

1) There are many influences that affect IIDP and its success.

2) One should distinguish and measure three types of consequences, at least:

(a) Innovations in management that enhance democracy and create possibilities for management of IIDP on the basis of the contributions of very many, or even all, of the co-workers in an organisation and in its outer network.

(b) Innovations resulting from idea management aimed at the collection of the inventions of many who work in- and outside the R&D circles, because this creates a pro-innovation VCEN and climate, and may result in many innovations yielding substantial benefits, including support to R&D.

3) The overall results that can be used for benchmarking via a balanced score card, etc.

The next set of influences to be measured, assessed, detected and hopefully, corrected by IIDP includes the enemies of innovation, such as (1) lengthy development times (for technological IIDP, especially the radical ones); (2) lack of co-ordination; (3) a risk-averse culture; (4) limited customer insight; (5) poor idea selection; (6) inadequate measurement tools; (7) dearth of ideas; and (8) marketing or communication failure. This survey has found that the greatest problem is due to inadequate measurement tools (McGregor, 2006).

3.2. The Metrics

In terms of the IIDP and the measurement of their outcomes, the traditional method is a rather simplistic measuring of inputs (McGregor, 2006) and the assumption that a comparatively high profit may be expected from them. However, sometimes there are other factors that hide the fruits of innovation (Henry, 2006). Patents are not a good yardstick; they are not innovations, but merely suggestions for or potential innovations, that is to say, only one of the many preconditions of innovation. Besides, it is a fact (e.g. in the US) that no more than one percent of patents actually become innovations. Neither can a reliable statement be based on R&D investments. They only cover technological inventions and their development into potential innovations, and do not take into account any other types of inventions or the necessary steps to turn an invention into an innovation, i.e. from an idea to its implementation in practice and its diffusion.

The number of genres of IIDP should constantly be increased (McGregor, 2006). If it is too low, as in the case of the traditional approach that takes into account the technological aspect alone, there is the risk of one-sidedness leading to failure due to oversights. Non-technological innovation, referred to as social innovation in IBM's document (IBM 2006), is at least equally crucial, because it creates preconditions under which IIDP may either flourish or be blocked (see: Mulej, 1981; Mulej, ed., 1984, 1987, 1994, etc; Rosenberg, Birdzell, 1986). The existence and efficiency of cross-functional collaboration is crucial for the same reasons (McGregor, 2006; Mulej, 1982).

In terms of outcomes, insight into current practices provides for a much less unified and clear picture: the span of measures reaches from zero to about eighty-five metrics, the sweet spot being found between 8 and 12 (McGregor, 2006). Given all the above-mentioned inputs, the input process needs much more attention than the outcomes. (See also: Germ Galic, 2004; Kroslin, 2004; Leder, 2004; etc.). Kroslin (2004) collected about eighty factors influencing IIDP, partly inside and partly outside the scope of influence of the organization at stake. Similar insights are provided by Innovation Score Boards and creditworthiness analysis (Knez Riedl, 1997 and later). They all deserve measuring or assessment, at least, if data are available and are suitable for being used as information rather than for creating misinformation. But one-sided measures/assessments provide for fictitious insights and wrong decisions, because they lack the requisite holism.

3.3 Requisite Holism, the Sustainable Enterprise and Happy People

Mulej & Kajzer's (1998) law of requisite holism provides a path between the two extremes in Figure 1; one should use the dialectical system of all crucial viewpoints (Mulej, 1974, and later). It is rarely that a single person will be able to attain the requisite holism as most tasks require a broader view. Thus, an interdisciplinary approach with creative co-operation of all crucial experts is what leads to the requisite holism in most cases. VCEN of interdependence, rather than of self-sufficiency and resulting independence, will achieve the requisite holism. (See: Mulej, 2007).

<-->		
Fictitious holism/realism (inside a single viewpoint)	Requisite holism/realism (a dialectical system of all essential viewpoints)	Total = real holism/realism (a system of all viewpoints)

Figure 1: The selected level of holism and realism in the consideration of the selected topic between the fictitious, requisite, and total holism and realism

In order to prevent avoidable cost of remediation of consequences of excessively one-sided decisions, the sustainable enterprise (SE) is the best solution in economic terms (Ecimovic et al., 2002; Knez-Riedl et al, 2001; Potocan and Mulej, 2007). An SE attains the highest level of requisite holism and contributes least to the destruction of the human capacity for survival. SEs are SEs because they are less one-sided than other enterprises and hence command with the most modern and comprehensive knowledge; they use VCEN enabling them to do no harm, or the least possible amount of harm, such as sustainable ethics resulting from sustainable development principles. This means, among other things, that the traditional economic criteria can no longer express reality, because they over-simplify; the very influential Mr. Forbes provides a good example of over-simplification (see quote in Mulej, N., 2006). The factoring-in of the criteria of sustainability diminishes the impression of the success of socio-economic development over the recent decades to a point where hardly any betterment of living standards remains (Bozicnik, 2007). SE criteria are more realistic, but perhaps not sufficiently so; criteria that reflect well-being may serve as well (Šarotar Žižek, 2012).

Diener and Seligman (2004) offer a promising new model for these criteria. It includes important non-economic predictors of the level of well-being, such as social capital, democratic governance, and human rights; all of them influence work satisfaction and productivity to a considerable extent. Supportive social relations are necessary for well-being. Well-being, in turn, will result in good social relationships with crucial economic policy implications. Desirable outcomes, even economic ones, often result from well-being, rather than the other way around. People high in well-being later earn higher incomes and perform better at work than others. They also have better relationships, are healthier, and attain longer lives. Thus, Diener and Seligman (2004) suggest measuring well-being with variables such as positive and negative emotions, commitment, purpose and meaning, optimism and trust, and life satisfaction.

Hornung (2006, p. 338) states that happiness is the permanent goal of humans and a holistic indicator of holistic well-being, well-functioning, and the physical, psychological, and social health of an individual. (We would call it a requisitely holistic indicator, total holism is impossible – Figure 1.)

This is important from the viewpoint of the model of four phases of economic development in Table 1 (Porter, 1990, quoted after Brglez, 1999; the authors added the resulting culture and phase 5).

Table 1: From misery via one-sided investment and innovation to affluence and from there to (hopefully) requisitely holistic creation

Development phase of basis of competitiveness	**Economic basis of the given development phase**	**Values – culture – ethics – norms typical of the given development phase**
1. Natural factors	Natural resources and cheap labor, hence poor life, for millennia	Modesty, solidarity, collectivism, tradition preferred to innovation
2. Investment in modern technology	Foreign investment, mostly; poor competitiveness in global markets; neglect of natural environment and health	Growing social differences based on property/inheriting, local competition, individualism, ambition to have more and become rich (in tangible property)
3. Innovation based on own capabilities	Nations/regions live on own progress, attaining growing competitiveness and standard of living	Social differences based on innovation, higher standard of living, global competition, ethics of interdependence, ambition to create
4. Affluence	People are rich, happy owners, no longer needing hard work for new progress	Complacency, consumerism, no more ambition to have more and hence to create
5. Requisitely holistic creation and social responsibility	Material wealth suffices; effort for it to be renewed and for spiritual wealth and healthy natural and social environment	Ethics of interdependence and social responsibility, hence ambition to create; diminishing of social differences so that only those caused by creation, including innovation, remain

It may mean that the affluence phase might be a dead end, if as a result people lose ambition for creation and thus become alienated from their human desire to create. People therefore need either a prolonged innovation phase based on a requisitely holistic IIDP rather than one-sided processes, or a new phase, a 5th

one, of creative happiness based on VCEN of interdependence and interdisciplinary creative co-operation that replaces the phase of (one-sided) affluence. These conditions are new and require new methods of measuring innovation and IIDP. The fifth phase that we all hope for requires requisitely holistic understanding of the current reality and of the role and importance of all humans in that reality, especially of the critical entities, such as commercial enterprises. This means that humans must use requisitely holistic thinking in their monitoring, perception, thinking, emotional and spiritual life, decision-making, and action for humankind to survive (For details see: Mulej, 1979; Mulej et al, 2000; Mulej, Kajzer, 1998; Potocan, 2000; and Potocan et al, 2005; Mulej, Dyck ed, 2014).

Understanding profits from measuring IIDP is the objective. The summarized range of influences over IIDP needs methods of measurement adapted to this complex reality. Where should one start? From management, of course, as it is the most influential beside governance.

4. Innovation as a Consequence of Management

The attributes of management discussed above are crucial for innovation and difficult to measure. Lester and Piore (2004) learned from surveys that quite a large number of practitioners consider IIDP an art based on intuition rather than on measurable factors. This is partly true as incremental innovations are easier to measure than the radical ones, and technological innovations are easier to measure than others.

Anywhere measurement of outcomes faces the same fade-down: measurable innovation is only the very last step of a demanding complex and complicated IIDP process from helping first ideas become registered rather than forgotten, followed by research, development, preparation of implementation in practice, and implementation leading to the final success in practice inside suggestion-authors' and owners' own organization or in any other market. Usually, there are many specifics to be addressed. Measures of innovation enhancement mostly used in connection with technological inventions are summarized below (Kavas et al, 2001, cited in Skafar, 2006, p. 32, adapted).

1) Market orientation of the IIDP;
2) Compliance of the IIDP with the general objectives of the organisation;
3) Efficient and effective selection of the invention-innovation project and its evaluation;
4) Efficient and effective project management and control;
5) Activation of sources of creative ideas;
6) Innovation-friendly organisational culture;
7) Engagement of influential individuals.

Types of technological IIDP to be measured include: research and development; acquiring of equipment for production of technologically new or improved products and production processes; acquiring intangible resources such as patents, licenses, trade-marks, models, know-how; industrial engineering, industrial design and testing production; training of employees for introduction of the technologically new or improved products or production processes; and marketing of new products (OECD, quoted in Skafar, 2006, p. 32).

These activities are easier to measure than other types of innovation: they are more tangible, but the list is far from complete. Hence, it might be interesting to assess how advanced a given organization may be in terms of the so-called innovative business and related culture (Mulej and co-authors, 1987; 1997, updated and enlarged – see below).

The bottom line of the innovative VCEN:
When managers and co-workers conclude: "This product / procedure / service is perfect", their next thought reads: "Let us start innovating it immediately, because our judgment of its perfection is based on criteria of 'so far', which customers and competitors may change any moment, causing them to withdraw their money from us".

This culture dictates innovative business:

1) In principle, every *cost/effort is unnecessary*. In reality it is so if we *work smarter*, not harder, and produce innovations.
2) Today, every product and process will sooner or later become *obsolete, with few exceptions such as diamond ring or other pieces of art*. That is why we must know their *life cycles*, do *research*, do *development* (connecting research results with daily requirements and practices), create other *inventions* and turn them into *innovations* as a new, useful / beneficial basis of survival, and diffuse them, on a *continuous* basis.
3) Survival and therefore both good and poor work is *everybody's business*. Nobody, neither the superiors nor the subordinates, is entitled to be *irresponsible* and to *oppose* or to *disregard* innovation in their own (work) life reality.
4) Therefore, let us continuously, *at all times and everywhere*, search for possible novelties! Only a small portion of them can become *inventions*. Some of them will be registered as *suggestions*. From some of them, by research and development, and/or connecting and developing concepts (Huston, Sakkab, 2006), sometimes something both usable and new might be created, *a potential innovation*. Customers will accept only a fragment of them as beneficial and worth paying for, hence making a benefit to both customers and suppliers, therefore deserving the name of *innovation*. They can be diffused, too, to support survival by business success (Rogers, 1995/2003).
5) Business policy and practice is *entirely innovation oriented*, not just a fragment of it.
6) *Results pay, not efforts*. Hence, let us work like the *clever* ones, not like fools. Diligent stupid humans are dangerous: they do it wrong mostly; so do clever bandits.
7) These six sentences no longer apply to the production departments of organisations only, but to *all activities and all parts of life* in all organisations, from a family to the United Nations.
8) The effort must be *broadly disseminated and permanent*, because the pressure of competitors is permanent.
9) For competitiveness, *quality* must be *systemic*, which is impossible without *continuous innovation*.
10) *Systemic quality* includes price, quality, flexibility, uniqueness and care for natural environment, all of them in interdependence, as a dialectical system.

Thus, there is hardly a chance to provide a unified short and simple set of measures for measuring or assessing innovation efforts in a simple way. It may even make more sense to provide measures as encouragement for enhancing, influencing IIDP and involvement of many/all members of the organisation in IIDP of any kind. This would make IIDP/innovation a normal daily routine so that VCEN of IIDP would prevail much more than it is doing at present. The entire world would then be much closer to becoming a generally innovative society; at present, only about 20 (twenty) percent of humankind might be said to belong to and benefit - at least directly - from such a society.

Hence, due to the managers' influence, the owners and governors should care for attributes of managers in organisations and their units. Rooke and Torbert (2005, p. 68) identified seven types of the managerial action logic. These include (1) opportunists, (2) diplomats, and (3) experts, mostly doing a good job on lower management levels and including 55% of all managers, while (4) achievers are 30%, and the last three types – (5) individualists, (6) strategists, and (7) alchemists account for no more than 15% and demonstrate the most consistent capacity to innovate and successfully transform an organisation. Some (ibid., 72), but not all (ibid., 76), leaders are able to transform from one action logic to another. Thus, people must be assessed, too, when one tries to identify causes of IIDP efficiency.

In addition, innovation may be viewed as a result achieved by knowledge workers. Their results tend to be implicit and hard to measure directly. Therefore, companies should concentrate on designing the processes that knowledge workers carry out rather than on measuring their performance (Economist, 2006, 13). Thus, co-workers are released from central control; the strongest glue holding them together is VCEN. In IBM (ibid, 16), they see its main attributes as:

1) Dedication to every client's success;
2) Innovation that matters, for our company and the world;
3) Trust and personal responsibility in all relationships.

Hence, measuring of IIDP success can be compared to measuring or assessing the above three attributes. Traditional measurement of efficiency does not seem to include them.

A specific feature of any organisation is its level of co-operation, both within its own walls and outside them. Thus, a proxy might be the level of VCEN of co-operation, as one is reminded by the experience quoted (ibid., 16). Another experience is from Nike (ibid., 18). Firms like Nike, employ few people directly. They become the so-called 'orchestrators' of a brand – this is what central officers are, frequently. Hence, the level of success in combining the VCEN of co-operation, interdependence and orchestration might be a proxy for managers' work on IIDP.

Another view of IIDP management is that of a case of knowledge management.

5. Knowledge Management and Innovation/IIDP

IIDP can be viewed as a case of knowledge management (KM): it transforms knowledge to new knowledge and the resulting yields. KM principles and practices can therefore be proxies for IIDP management assessment. They include (Palacios-Marqués, Garragós-Simón, 2005, 361):

1) Orientation towards the development, transfer and protection of knowledge;
2) Continuous learning within the organisation;
3) Understanding of an organisation as a global system (i.e. complex entity);
4) Development of an innovative culture that encourages R&D projects (and others, we add), based on individuals;
5) Approach to competence development and management based on competencies.

To these, one more principle should be added (Basadur, Gelade, 2006, 45); organisational creativity and innovation/IIDP can be integrated with KM and organisational learning in a single framework that combines the apprehension of knowledge with the creative utilization of such knowledge.

This framework allows organisation to do three things:

1) Detect errors and implement changes to restore or improve routines;
2) Make sense of sudden unexpected events and crises and convert them into opportunities for innovation/IIDP; and
3) Anticipate and seek out new information, and emerging opportunities to develop new products, services, and routines.

The soft factors, such as culture, leadership, and motivation, are important for the success of IIDP, but not enough to implement transformation projects, because they do not directly influence the outcomes of many change programs (Sirkin et al., 2005, 110). Important hard factors are easier to measure, their importance is easier to communicate, and they are easier to influence quickly (ibid., 110):

1) The time necessary to complete the project;
2) The number of people required to execute it; and
3) The financial results expected.
4) Four hard factors the same authors have detected, empirically, include (ibid., p. 110):
5) Project duration;
6) Performance integrity (i.e. project team capability);
7) Commitment of both senior executives and the staff whom the change will affect most; and
8) The additional effort that employees must make to cope with the change.

The authors refer to these factors as DICE.

Yet another story is the early phase of IIDP, called idea management. It also covers the Suggestion System or Submit Your Idea System or Reinventing as a part of IIDP.

6. Idea Management and Measuring

The activation of many or even all co-workers as potential inventors and innovators can be assessed by measures included in Idea Management (IM), as follows.

The set of quantitative output indicators proposed to measure the efficiency of the IM program consists of the following (Fatur, Likar, 2008;), which would mostly address the tactics and operation phases of the business process:

1) SUG/EMP measures the rate of the employees participating in the program as suggestion providers. This indicator relates to the ability of the IM to attract a critical mass of coworkers/employees into the program (its ability of popularisation – VCEN, climate, promotion, incentives etc.).
2) SBM/SUG indicates the average number of submissions per suggestion provider. Attracting a person to submit his/her first suggestion requires very different management methods than retaining him/her in the program (among them prompt and fair treatment of suggestions).
3) APP/SBM is the number of approved suggestions among all submitted. Approved suggestions are normally the ones which prove useful to the company. High APP/SBM value indicates the management's ability to focus the employees on the areas that need improvement.
4) IMP/APP indicates the number of implemented ideas among all approved. Usually, implementation of suggestions is outside the IM domain. Therefore, this indicator shows the ability to motivate the line managers (e.g. maintenance staff) to put suggestions into practice (to turn inventions into innovations).
5) NSV/IMP stands for the calculated net benefits per implemented suggestion.

A multiplication of the five output indicators results in an aggregate indicator NSV/EM*P* (net benefit per employee). A high *NSV/EMP* shows that the IM process has managed to attract a large portion of all the employees as suggestion providers, many suggestion providers have submitted several ideas, many of them have been approved, and many approved ones have been implemented, yielding high financial benefits. Therefore, reaching a high *NSV/EMP* value should be the primary goal of any idea manager.

To make the job easier, one must consider the following practical experiences:

1) Definition of vision, mission etc. must not be a formality on paper, but a tangible reality; otherwise managers depend on IIDP/innovation fanatics only.
2) Definition of suggestions must be very clear, published and practiced all the time.
3) Actions to collect criticism and ideas must become a campaign 2-4 times a year, if they do not take place on a weekly basis; higher managers ask lower managers to collect criticism (on an anonymous basis), as well as suggestions (on a non-anonymous basis, if authors so wish, or with a secret sign).
4) In every organisation, sometimes similar criticism and suggestions are presented at the same time. Therefore, the invention/innovation office must run a computer-supported IM information system. It must be available to all users of the organisation's information system and apply library methodology of access, with various viewpoints and their synergies in mind.
5) Transfer of potential innovation from one unit to another may take place, but there may be a lack of knowledge about each other and of trust in each other. Enthusiasm and a knowledge-base for transfer must be built in advance, including the hands-on workers involved.
6) Some suggestions may require rather a high investment, which may be impossible in the short run. Such cases should be put on hold and made to wait, while the authors should immediately receive their innovation awards or at least a part of the estimated expected reward compensation.
7) Sometimes delays in implementation of promising potential innovations occur. For instance, people from the engineering office people may be overburdened or lack interest. In such a case, they should be made co-authors, including receiving a part of the innovation award. Sometimes it is good to sell the idea to a business partner and have them implement it.
8) Innovating the information/communication system is usually a complex

endeavour. It requires thorough knowledge of all processes to be tackled. Adaptation of both the processes and the ICT must be taken care of very early in the preparation of the contract.

9) Some potential innovations face obstacles. One should put them on hold and take a look at them quite frequently, while carefully waiting for a window of opportunity to apply them. According to the 'open innovation' concept, they might become the basis of a spin-off company, or sold (Chessbrough, 2003). Hence, shelfing and waiting for a 'better opportunity', may not be better than marketing the idea that otherwise might become obsolete since there are many researcher and competitors out there.
10) Sometimes employees are jealous of their more creative colleagues and their awards, although these awards have been fully deserved. The solution is team work, i.e., including them as persons with some knowledge and experience.
11) And so on.

These results are the easiest to measure – see Table 2. Of course, not only the impact measurement per phases of IIDP, but also the final outcomes matter.

Table 2: Measurement of Idea Management results – a case International research by Pisk d.o.o.: MI GIMB (Global Idea Management Benchmarking) http://www.pisk.si/download/gimb_rezultati.pdf, some Slovenian companies.

Year	**Co.**	**Sug/Emp**	**Sbm/Sug**	**App/Sbm**	**Imp/App**	**Nsv/Imp (€)**
2004	Unior	6 %	2	56 %	76 %	1728
2005	Unior	8 %	2,1	47 %	99 %	1750
2006	Unior	11 %	1,68	49 %	88 %	3078

Year	**Enterprise**	**Sug/Emp**	**Sbm/Sug**	**App/Sbm**	**Imp/ App**	**Nsv/Imp (€)**
2003	Vogt electronic d.o.o.	64 %	2,8		89,3 %	289,14
	BSH Hisni aparati d.o.o.	22 %	1,23		40,5 %	843,54
	LIV Postojna d. d.	54 %	0,58		34,8 %	2134,61
	Cimos d.d.	23%	0,47		95,3 %	445,94

	Comet d.d.	24 %	0,47		97,0 %	725,56
	Kolektor PRO d.o.o.	19 %	0,44		99,4 %	1008,23
	Iskraemeco d.d.	12 %	0,44			187, 25
	Grammer Automotive Slovenija d.o.o.	14 %	0,40		66,0 %	393, 94
	Arcont d.d.	13 %	0,37		55,8 %	582, 19
	Luka Koper d.d.	14 %	0,33			955,06
	IMP Klimat d.d. Ljubljana	22 %	0,32		50,0 %	3000
	Domel d.d.	49 %	0,31			982,72
	Iskra avtoelektrika d.d.	17 %	0,29		68,2 %	2370,69
	Droga d.d.	12 %	0,29		42,9 %	833,33
	Trimo d.d.	21 %	0,20			7405,06
	Expo biro d.o.o.	7 %	0,20			
	Stikala	7 %	0,15		77,8 %	
	ETI Elektroelement d.d. Izlake	6 %	0,12		79,2 %	
	Zito Gorenjka d.d.	1 %	0,01		100 %	18750,5
2004	Acroni d.o.o.		0,41			3873,14
	Comet d.d.	94 %	0,36			424, 53
	Grammer Automotive Slovenija d.o.o.	79 %	1,01			213,68
	Iskra avtoelektrika d.d.	96 %	0,23			2759,66
	Luka Koper d.d.	100 %	0,36			738,39
	Trimo d.d.	94 %	0,64			11333,33
	Vogt electronic d.o.o.	94 %	2,87			218,28

7. The Overall Results of IIDP

For all organisations, successful IIDP management and hence the innovation measurement issue is critical. This is especially true for the so-called intelligent organisations, which are, at the same time, aware of opportunities and risks, stemming from IIDP, want to base their strategic decisions on reliable information about IIDP, and make it available as soon as possible. Numerous partial data are not as expressive as ratios and systems of ratios, where several relevant categories are interlinked.

Partaking of a multi-dimensional nature, IIDP is very complex. This fact makes the issue of innovation measurement even more difficult. Additionally, the typology of innovation is becoming broader. Besides product and market innovations, other types attract attention, e.g. process innovation, managerial innovation, service innovation, and social innovation (Knez-Riedl, Hrast, 2005). The subtle nature of innovation is coming into force, demanding well-thought-out measures.

Regarding innovation metrics, four generations can be distinguished (Milbergs, Vorontas, downloaded in 2006) – Table 3.

Among approaches that try to take into account the complex nature of innovation, the so-called *Innovation Framework* monitors innovation from three viewpoints: resources, capabilities, and leadership. The emphasis regarding resources is put on balancing allocation of capital, work, and time. In the context of capabilities, the organisational VCEN are especially important. Equally, leaders' support of and involvement with promotion and dissemination of innovation goals is critical, as well. All three aspects are linked by processes that comprise organisational structures (innovation incubators, markets, funds, and incentives). Each aspect is illustrated by selected, understandable indicators that encourage people to strive for the innovation goals.

Table 3: Four generations of innovation metrics

Source: Milbergs and Vonortas, 2004

1st generation Investment indicators (1950-60)	**2nd generation Result indicators (1970-80)**	**3rd generation Innovation indicators (1990)**	**4th generation Process indicators (Since 2000)**
• R&D expansion • Science &	• Patents • Publications • Products	• Studies • Indexes • Benchmarking of	• Knowledge • Intangibles • Networks

Technology staff • Capital • Technical intensity	• Change in quality	innovation capability	• Demand • Clusters • Managerial techniques • Risk/return • System dynamics

Resources:

(1) Inputs: capital, talent, and time

(a) Percentage of capital, invested in innovations

(b) Number of entrepreneurial employees, previous entrepreneurs

(c) Percentage of working time devoted to innovation projects

(2) Outputs: return on investment

(a) Number of new products, services and jobs launched in the past year

(b) Percentage of revenues of products and services introduced in the past three years

(c) The change of market value of a firm in the past year, in comparison with the change of the whole industry market value

Capabilities:

(1) Inputs: conditions

a) Percentage of employees for whom innovations are the core business goal

b) Percentage of employees trained for innovating

c) Number of available innovation tools and methodologies

(2) Outputs: reengineering

a) New professional knowledge development

b) New strategic possibilities and growth basis

c) New markets, won in the past year

Leadership:

a) Percentage of leadership time devoted to strategic innovations

b) Percentage of managers trained with regard to the concepts and tools required for innovation

c) Frequency of re-evaluation of the key business areas

Processes

a) Number of suggestions in the past three, six and twelve months
b) Successful ideas/suggested ideas
c) Number of trials and successful ideas in due course
d) Average period from idea to launching/commercialisation

Since Kaplan and Norton (1992) launched their *Balanced Scorecard* (*BSC*), the modifications of measurement appear also with regard to innovation. There are several versions of the so-called *Innovation Scorecard.* Like the original idea underlying *BSC,* all these different models strive for a more holistic management of innovation/IIDP. For instance, *Peter S. Cohan & Associates* (www.businessinnovationinsider.com) are tackling the goal of *return on investment* from the following aspects:

a) Entrepreneurial leadership (e.g. percentage of employees who are able to improve competitiveness),
b) Allocation of resources,
c) Technology (e.g. number of individuals responsible for monitoring new technologies), and
d) Product development (e.g. how cross-sectional and cross functional collaboration contributes to new products and their commercialisation, how many working groups report quick response from customers).

The central goal – return on investment – is measured by the following measures:

a) Profit and revenue per employee, in comparison with competitors;
b) Internal rate of return on investment in innovation;
c) Profit from new products and processes in comparison with investments in innovation.

Several other indicators and ratios have been developed for measuring innovations. *Medori (2000)* proposed several ratios, especially emphasising time required from design to final commercialisation. *Müller* and *von Thiesen (2001)* developed a catalogue of ratios related to so-called e-profit. In this context, the emphasis is put on ratios related to innovation/IIDP, education, knowledge, and e-business.

Besides individual ratios and ratio systems, some aggregates combine selected ratios. E.g. *Dow Corning Innovation Index* takes into account the generation of ideas, the percentage of employees and revenue growth.

The variety of measures mentioned tends to form a useful toolkit for:

a) Increasing the transparency of innovation/IIDP,
b) Enabling planning, decision-making, realisation, analysis and controlling of innovation/ IIDP (all of these both on a long and a short term basis),
c) Enabling time comparison and benchmarking,
d) Encouraging improvements in performance,
e) Helping to manage risks related to innovations, and
f) Rewarding of innovators.

In the author's experience, these data tend to be kept as business secrets.

8. Diffusion as an Unavoidable Phase of IIDP

Last, but not least, there is need for the diffusion of inventions that have become potential innovations and/or have started to become innovations, for the highest potential benefit to be achieved. This also is a very complex phase and topic. Diffusion consists of many activities that also require measurement. If they are not efficient, the IIDP has no effective end. For details, see (Rogers, 2003, and earlier editions).

Conclusion

Innovation as an outcome can be attained in any market only as the final phase of IIDP. This fact makes all organisations – bigger and smaller, tangible and virtual, enterprises and others, owned by individuals, families, shareholders or governments, local and international, etc. – equal. IBM (2006) and McGregor (2006) published results of surveys showing that more managers, although still too few for an innovative society, show awareness that innovation is necessary than admit personal involvement in IIDP. Therefore, it is not good enough to measure end results of this process only and leave aside the impacting factors. Here, the authors have collected some reminders and other potential supports to IIDP management. However, measuring them is very poorly described in literature.

The contemporary 'open innovation' concept (Chesbrough, 2003; Chesbrough et al, 2006; Fujs, Mulej, 1993; Huston, Sakkab, 2006) or '(corporate) social responsibility' (Hrast et al, editors, 2006, 2007, 2008) may make measuring of enterprise efficiency even more complex. These issues will be left aside for further research.

References

Books

Barabba, V. P. (2004) *Surviving Transformation. Lessons from GM's Surprising Turnaround.* Oxford Press: Oxford.

Belak, J. (2003) *Integralni management in razvoj podjetja.*

Bozicnik, S. (2007) *Dialekticno sistemski model inoviranja krmiljenja sonaravnega razvoja cestnega prometa (Ph.D. Diss.).*

Brglez, J. (1999) *Razvojni potenciali majhnih gospodarstev v razmerah evropskega integracijskega procesa.*

Chesbrough, H. (2003) *Open Innovation. The New Imperative for Creating and Profiting from Technology.* Harvard Business School Press Books.

Chesbrough, H., Vanhaverbeke, W. and West, J. (2006) *Open Innovation: Researching a New Paradigm.* Oxford University Press.

Ecimovic, T., Mulej, M. and Mayur, R. (eds.) (2002) *System Thinking and Climate Change System (Against a big "Tragedy of Commons" of All of Us).*

Feucht, H. (1995) *Implementierung von Technologiestrategien.* Frankfurt Main.

Florida, R. (2005) *Vzpon ustvarjalnega razreda (Translation to Slovenian, original 2002).*

Fujimoto, T. (2007) Competing to Be Really, Really Good: The Behind-The-

Scenes Drama of Capability-Building Competition in the Automobile Industry. cTokyo: LTCB International Library.

Germ Galic, B. (2003) *Dialekticni sistem kazalnikov inoviranja in kakovosti poslovanja (M.A. Thesis).*

Gu, J.A, Nakamori, Y., Wang, Zh., Tang, X., eds (2006) Towards knowledge synthesis and creation. Proceedings of the Seventh International Symposium on Knowledge and Systems Sciences, Beijing, China, September 22-25, 2006, (Lecture Notes in Decision Sciences, 8). Hong Kong etc.: Global-Link Publisher.

Gustin, N. (2007) *Aktiviranje ustvarjalnosti za vecjo inovativnost poslovanja podjetja.* M.A. Thesis. Maribor: University of Maribor, Faculty of Economics and Business.

Hrast, A., Mulej, M., Knez-Riedl, J., editors (2006) *Druzbena odgovornost in izzivi casa 2006. Zbornik prispevkov.* Maribor: IRDO. On CD (abstracts in book, bilingual).

Hrast, A., Mulej, M., Knez-Riedl, J., editors (2007) *Druzbena odgovornost 2007. Proceedings of the 2nd IRDO Conference on Social responsibility.* IRDO Institut za razvoj druzbene odgovornosti. Maribor. On CD (abstracts in book, bilingual, several texts in English).

Hrast, A., Mulej, M., editors (2008) *Druzbena odgovornost in izzivi casa 2008. Zbornik prispevkov.* Maribor: IRDO. On CD (abstracts in book, bilingual).

Kroslin, T. (2004) *Vpliv dejavnikov invencijsko-inovacijskega potenciala na uspesnost podjetij.* Maribor: University of Maribor, Faculty of Economics and Business.

Leder, B. (2004) *Inoviranje trzenja turizma na slovenskem podezelju.* M. A. thesis. Maribor: University of Maribor, Faculty of Economics and Business.

Lester, R. K., M. J. Piore (2004) *Innovation - The Missing Dimension.* Cambridge, Ma, London: Harvard Business Press.

Mesner Andolsek, D. (1995) *Vpliv kulture na organizacijsko strukturo.* Ljubljana: Gospodarski vestnik.

Mulej, M., Dyck, R., editors (forthcoming in 2014): *Social Responsibility beyond Neoliberalism and Charity.* BenthamScience, Shirjah, UAE

Mulej, M., Dyck, R., editors (forthcoming in 2014) Social Responsibility beyond Neoliberalism and Charity. BenthamScience, Shirjah, UAE.

Mulej, M. (1975) *Dialekticna teorija sistemov* (In Slovenian). Ljubljana: University of Ljubljana, Faculty of Sport. No ISBN available. Unpublished lecture.

Mulej, M. (1981) *O novem jugoslovanskem modelu druzbene integracije.* Maribor: Zalozba Obzorja.

Mulej, M. (1979) *Ustvarjalno delo in dialekticna teorija sistemov.* Celje: Razvojni center Celje.

Mulej, M. (2006) *Absorbcijska sposobnost tranzicijskih manjsih podjetij za prenos invencij, vednosti in znanja iz univerz in institutov.* Koper: University of Primorska, Faculty of Management.

Mulej, M., Devetak, G., Drozg, F., Fers, M., Hudnik, M., Kajzer, S., Kavcic, B., Kejzar, I., Kralj, J., Milfelner, R., Mozina, S., Paluc, C., Pirc, V., Pretnar, B., Repovz, L., Rus, V., Sencar, P., Tratnik, G. (1987) *Inovativno poslovanje.* Ljubljana: Gospodarski vestnik.

Mulej, M., de Zeeuw, G., Espejo, R., Flood, R., Jackson, M., Kajzer, S., Mingers, J., Rafolt, B., Rebernik, M., Suojanen, W., Thornton, P., Ursic, D. (1992) *Teorije sistemov.* Maribor: University of Maribor, Faculty of Economics and Business.

Mulej, M., and coauthors Hyaeverinen, L., Jurse, K., Rafolt, B., Rebernik, M., Sedevcic, M., Ursic, D. (1994, reprinted also in 2007) *Inovacijski management. I. knjiga Inoviranje managementa.* Maribor: University of Maribor, Faculty of Economics and Business.

Mulej, M., Zenko, Z. (2004) *Introduction to Systems Thinking with Application to Invention and Innovation Management.* Maribor: Management Forum.

Mulej, M., ed. (1997) *Obvladovanje inovacij in kakovosti.* University of Maribor, Faculty of Economics and Business. Maribor.

Mulej, M., Espejo, R., Jackson, M., Mingers, J., Mlakar, P., Mulej, N., Potocan, V., Rebernik, M., Rosicky, A., Schiemenz, B., Umpleby, S., Ursic, D., Vallèè, R. (2000) *Dialekticna in druge mehkosistemske teorije (podlaga za celovitost in uspeh managementa).* Maribor: University of Maribor, Faculty of Economics and Business.

Mulej, M., Zenko, Z. (2004) *Dialekticna teorija sistemov in invencijsko-inovacijski management. (Kratek prikaz).* Maribor: Management Forum.

Mulej, M., et al (2013) *Dialectical Systems Thinking and the Law of Requisite Holism concerning Innovation.* Litchfield Park, Arizona: Emergent Publications.

Müller, A., and von Thiesen, L. (2001) e-Profit Controlling-Instrumente *für erfolgreiches e-Business*, Freiburg: Rudolf Haufe Verlag.

Peters, T. (1997) *The Circle of Innovation.* New York: Knopf.

Reich, R. (1984) *The Next American Frontier.* New York: Penguin Books.

Rogers, E. (1995) *Diffusion of Innovation.* The 4th edition. New York: The Free Press.

Rosenberg, N., Birdzell, L. E. (1986) *The Past. How the West Grew Rich.* New York. Basic Books.

Šarotar Žižek, S. (2012) Vpliv psihičnega dobrega počutja na temelju zadostne in potrebne osebne celovitosti zaposlenega na uspešnost organizacije. Maribor:

UM, EPF.

Skafar, B. (2006) *Inovativnost kot pogoj za poslovno odlicnost v komunalnem podjetju*. Ph.D. Diss. Maribor: University of Maribor, Faculty of Economics and Business.

Thorpe, S. (2003) *Vsak je lahko Einstein. Krsite pravila in odkrijte svojo skrito genialnost*. Ljubljana: Zalozba Mladinska knjiga.

Zenko, Z. (1999) *Comparative Analysis of Management Models of Japan, USA, and Europe*. PhD. Diss. Maribor: University of Maribor, Faculty of Economics and Business.

Journal Articles:

Basadur, M., and Gelade, G. A. (2006) The Role of Knowledge Management in the Innovation Process. *Creativity and Innovation Management*, 15, 1, 45-62.

Diener, E. and Seligman, M. E. P. (2004) Beyond Money. Toward an Economy of Well-Being. *Psychological Science in the Public Interest,* Volume 5, No 1, 1-31.

Economist (author not mentioned) (2006) The new organization. A survey of the company. In *The Economist*, 21 January, 1-20.

Fatur, P. and Likar, B. (2008) Development of performance measurement methodology for idea management. *International Journal of Innovation and Learning*, accepted for publication.

Fujs, E., Mulej, M. (1993) 21 new attested products after five years of restructuring: Primat Maribor. *Public Enterprise*, 14, March-June, 142-147.

Henry, D. (2006) Profit Margins. Creativity Pays. Here's How Much. *BusinessWeek*, April 24, 76.

Hornung, B. R. (2006) Happiness and the pursuit of happiness. A sociocybernetic approach. *Kybernetes*, 35, 3/4, 323-346.

Huston, L., Sakkab, N. (2006) Connect and Develop. Inside Procter & Gamble's New Model for Innovation. *Harvard Business Review*, March 2006, 1-9.

Kaplan, R. S. and Norton, D. P. (1992) The Balanced Scorecard – Measures that Drive Performance, Harvard Business Review, January – February, 1992, 71 – 79.

McGregor, J., et al. (2006) The World's Most Innovative Companies. *BusinessWeek*, April 24, 63-74.

Mulej, M. (1982) Dialekticno sistemsko programiranje delovnih procesov – metodologija USOMID. *Nase gospodarstvo*, 28, 3, 206-209.

Mulej, M. (2007) Systems theory – a worldview and/or a methodology aimed at requisite holism/realism of humans' thinking, decisions and action. *Systems*

Research and Behavioral Science, 24, 3, 347-357.

Mulej, M. (2006) Steve Forbes, chairman and editor-in-chief of Forbes. *Marketing magazin*, No 307, November, 26-27.

Mulej, M., ed. (1984) *Vgrajevanje inventive v politiko in prakso OZD. 5. PODIM.* Proceedings in *Nase gospodarstvo*, 30, 1-2.

Palacios-Marqués, D., and Garrigós-Simón, F., J. (2005) A measurement scale for knowledge management in the biotechnology and telecommunications industries. *International Journal of Technology Management*, 31, No 3-4, 358-374.

Potocan, V. (1997) New perspectives on business decision making. *Management: Journal of contemporary management issues*, 2, 2, 13-24.

Potocan, V., Mulej, M. (2007) Ethics of a Sustainable Enterprise and the Need for it. *Systemic Practice and Action Research*, 2007, vol. 20, no. 2, 127-140.

Potocan, V., Mulej, K., and Kajzer, S. (2005) Business Cybernetics. *Kybernetes*, 34, 9/10, 1496-1516.

Rooke, D., Torbert, W. R. (2005) 7 Transformations of Leadership. *Harvard Business Review*, April, 67-76.

Sirkin, H. L., Keenan, P., and Jackson, A. (2005) The hard side of change management. *Harvard Business Review*, October, 109-118.

Contributions to books:

Dewulf, S. (2006) Directed Variations® Systematic Innovation in the Established Companies. In: Rebernik, M., Mulej, M., Rus, M., Kroslin, T., eds.: *Cooperation between the economic, business and governmental spheres: Mechanisms and levers. Proceedings of the 26 PODIM Conference on Entrepreneurship, Innovation, and Management.* Maribor: University of Maribor, Faculty of Economics and Business. Institute for Entrepreneurship and Small Business Management, 53-67.

Kettula, T. (2005) The Role of Finnish Science Parks and Experiences on Building regional Innovation Systems. In: Rebernik, M., Mulej, M., Rus, M., Kroslin, T., eds.: *Ustvarjanje okolja za prenos inovacij / Shaping the Environment for Innovation Transfer. Proceedings of the 25 PODIM Conference on Entrepreneurship, Innovation, and Management.* Maribor: University of Maribor, Faculty of Economics and Business. Institute for Entrepreneurship and Small Business Management, 71-83.

Knez-Riedl, J. (1997) Creditworthiness of Potential Business Partners as a Precondition for Networking of Groups and Enterprises. In: Brandt, D., ed.: *Automated Systems Based on Human Skills.* IFAC Symposium. Preprints. Ljubljana: University of Technology Aachen, and J. Stefan Institute, for IFAC.

Knez-Riedl, J., Mulej, M. and Zenko, Z. (2001) Approaching sustainable enterprise. In: Lasker, G. E., Hiwaki, K. (Eds.): *Sustainable development and global community.* Baden Baden: International Institute for Advanced Studies in Systems Research and Cybernetics.

Knez-Riedl, J. and Hrast, A. (2005) Innovation in the context of corporate social responsibility (CSR). In: Bulz, N. et al. (Eds.): *Proceedings of The WOSC 13th International Congress of Cybernetics and Systems, 6-10 July, 2005* Maribor: Faculty of Economics and Business, vol. 6, 45-54.

Medori, D. Innovation Performance Measurement in World Class Manufacturers: The State of Play. In: Proceedings of the 5th International Conference on Linking Systems Thinking, Innovation, Quality, Entrepreneurship and Environment STIQE'00, 2000. Maribor: University of Maribor, Faculty of Economics and Business. 163-161.

Mulej, M. (2006b) Requisitely Holistic Management of the Invention-Innovation Process as a Specific Case of Knowledge Management. Invited paper to: Gu, J. et al, editors: The 7th International Symposium on Knowledge and Systems Science "Towards Knowledge Synthesis and Creation", in Beijing, September 22-25.

Mulej, M., and Kajzer, S.: Self-Transformation and Transition from a Pre-industrial to a Contemporary Economy and Society; in Dyck, Mulej 1998, referenced here, 325-332.

Mulej, M., Kajzer, S. (1998) Ethics of Interdependence and the Law of Requisite Holism. In: Rebernik, M., Mulej, M., eds.: STIQE '98. Proceedings of the 4th International Conference on Linking Systems Thinking, Innovation, Quality, Entrepreneurship and Environment. Maribor: Institute for Entrepreneurship and Small Business Management, at Faculty of Economics and Business, University of Maribor, and Slovenian Society for Systems Research, 129-140.

Mulej, M., Mulej, N. (2006) Innovation and/by Systemic Thinking by Synergy of Methodologies "Six Thinking Hats" and "USOMID". In: Trappl, R., ed.: *Cybernetics and Systems Research 2006.* Vienna. Austrian Society for Cybernetic Studies, 416-421.

Parra Luna, F. (2007) An Axiological Concept of Organisational Efficiency: a Measure (Toward Measuring the Efficiency of Firms). In: Wilby, J. editor (2007): *Proceedings of the 51st Annual Conference of the International Society for Systems Sciences: Integrating Systems Sciences: Systems Thinking, Modeling and Practice.* Tokyo. Tokyo: Tokyo Institute of Technology, and ISSS.

Sato, Ch., Kumagai, S., Tsukuda, J., Numata, J. (2005) Innovation in Business Practice Unit – Indices of Innovation and Innovation Engineering. In: Rebernik, M., Mulej, M., Rus, M., Kroslin, T., eds.: *Ustvarjanje okolja za prenos inovacij/Shaping the Environment for Innovation Transfer. Proceedings of the 25 PODIM Conference on Entrepreneurship, Innovation, and Management.* Maribor: University of Maribor, Faculty of Economics and Business. Institute for Entrepreneurship and Small Business Management, 187-199.

Internet articles:

Milbergs, E., Vonortas, N.: Innovation Metrics: Measurement to Insight. Paper prepared for National Innovation Initiative 21st Century Innovation Working Group.
http://innovate.typepad.com/innovation/files/innovation_metrics_issue_paper_1.4%20Dec%202005.pdf [Available: 5.8.2008]

Official documents:

EU (1995) *Green Paper on Innovation. Draft. December 2005.* European Commission.
http://europa.eu/documents/comm/green_papers/pdf/com95_688_en.pdf [Available: 5. 8. 2008] (EU document number: COM (95) 688 final, pages: 131, ID code: 1218).

EU (2000) *Communication from the Commission to the Council and the European Parliament: Innovation in a knowledge-driven economy.* Commission of the European Communities, Brussels, xxx COM (2000) 567 final.

Work in process:

Bornsek, G. (in process) *Dialekticni sistem meril in standardov uspesnosti inoviranja v tranzicijskih podjetjih.* Ph.D. Thesis. Maribor: University of Maribor, Faculty of Economics and Business.

Kokol, A. (in process) *Posebnosti managementa RR v srednje velikem proizvodnem podjetju v tranzicijskem gospodarstvu.* Ph.D. Thesis. Maribor: University of Maribor, Faculty of Economics and Business.

6

The Environmental Efficiency of Air Transport: can it be Measured in Today´s World of Complexity and Chaos?

By Antonio Sanchez-Sucar

Abstract

Air Transport (AT) is a realm where conflicts of interest often play a key role in driving the economic and political conflict over issues of sustainability, not only because it implies different things to different people but also because even if a shared meaning of sustainability can be reached, it is not always clear how strategies and goals can be attained. The purpose of this paper is to review the policy and commitment of AT with regard to sustainability and to propose some indicators to measure its efficiency. The theoretical frame used is based on the fact that efficiency - defined as the ability of an organisation to maximize its desired outputs with respect to the inputs - is something that can and must be quantitatively measured. And this implies: a) the consideration of an axiological frame and b) an effective and consistent metrics system. The System Theory is applied for this endeavour in order to show the different subsystems and their interrelations. Some sustainability indicators are proposed within the general axiological frame of the "Reference Pattern of Values", particularly those related to "Nature Conservation" and "Freedom of Mobility".

1. Introduction

The origin of environmentalism can be traced back for more than a century, but it reached social and scientific maturity in the second half of the 20th century, when a more precise, but still general concept of sustainable development was formulated. However, a clear definition of such a concept has proved to be difficult; in the late 80´s, there were still multiple interpretations of it.

As of today, two main streams of sustainable development can be identified: economic and societal (anthropocentric). The former claims that economic growth as a cause of environmental deterioration must be limited to the carrying capacity of the earth, and focus on the efficiency of the relevant processes, while the latter refers to a “desirable condition for mankind” which, while indicating directions for change, provides a less explicit set of instruments to measure and govern this change.

Most authors agree that if sustainability is to be made operational, appropriate methods and indicators should be available for its assessment. This is particularly true where one is obliged to deal effectively with complex problems derived from the interactions of complex systems. Taking into account the organisational complexity and the existing diverse scenarios affecting the activities of AT, it seems to be appropriate to use a methodology based on Systems Theory in such a way that efficiency is measured as an aggregation of the efficiency of different subsystems, defined as output/input ratios (Parra Luna, Francisco, 1982). Some authors define these ratios as productivity indicators. In our case, this would apply to the economic (Y_E/X) indicators, where Y_E is production (economical) output and X stands for system input (Fig.1).

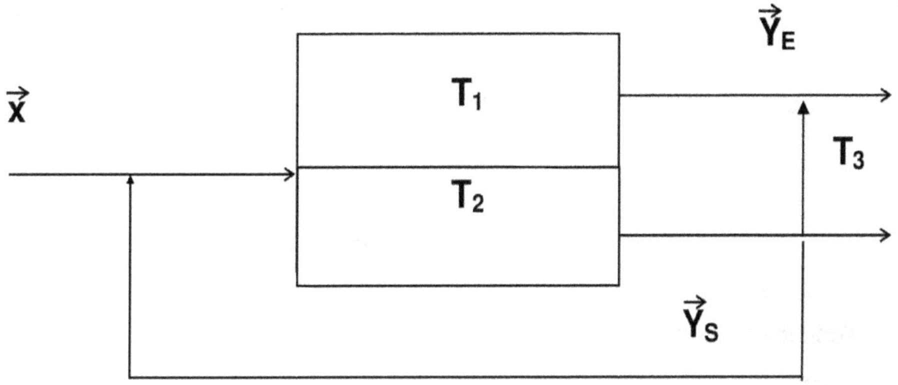

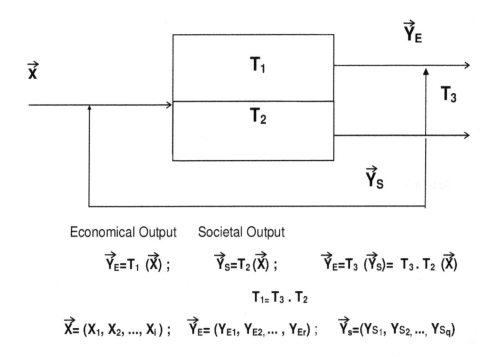

Economical Output Societal Output

$$\vec{Y}_E = T_1(\vec{X}); \qquad \vec{Y}_S = T_2(\vec{X}); \qquad \vec{Y}_E = T_3(\vec{Y}_S) = T_3 . T_2(\vec{X})$$

$$T_1 = T_3 . T_2$$

$$\vec{X} = (X_1, X_2, ..., X_i); \quad \vec{Y}_E = (Y_{E1}, Y_{E2}, ..., Y_{Er}); \quad \vec{Y}_s = (Y_{S1}, Y_{S2}, ..., Y_{Sq})$$

Figure 1: Economical and Societal Outputs

All the same, organisations of the business sector additionally produce "societal goods" (Y_S) that can be subjected to measurement through societal indicators of the form Y_S /X or Y_S/Y_E. Their environmental impact can be considered a societal good in terms of social "disbenefit" to be minimised (hence affected by a (-) sign). However, quantitative measurements alone cannot define what is "good" or "bad", and a reasonable consensus among all participating actors of a system where different values and interests coexist must be reached. Other uncertainties derive from the fact that there is a set of widespread common protocols of a global nature. The availability of accepted and shared protocols is a condition of progress, but that is not the end of the road. (Upham, Paul et al, 2003). The question is: are efficiency indicators available for AT? Have they the potential of making a global, homogeneous way of sustainability measurement possible? The AT considered in this paper refers only to Civil Aviation. We have not included military aviation, although its goals are widely shared. Boeing and EADS, the two major aircraft manufacturers worldwide have military branches and it is usual for new technology to be first applied within the framework of military projects.

Among the various definitions of sustainability existing in literature, it seems appropriate to use one that directly addresses the concept of efficiency. Following Daly (1991):

a) Technological progress for sustainable development should be efficiency-increasing, rather than throughput (raw materials)-increasing.Deterioration should not exceed regeneration rates, and wastes and emissions should not exceed the renewable assimilative capacity of the environment.
b) Non-renewable resources should be exploited, but at a rate that does not exceed the creation of renewable substitutes.

2. AT today

After the development experienced during the past 50 years, today's AT is a complex system that not only affects (and is affected by) the global economy, but also fosters the perception of the public towards easier and global mobility. During this period, the traffic volume in the USA (in Available Tonne Miles, ATM) has increased by a factor of 30[9], with the same trip cost in constant dollars. The worldwide passenger traffic (in Revenue Passenger Kilometers, RPK) has grown since 1960 by an average rate of 9%, that is to say 2,4 times more than the GDP (ICAO. Yearly Report, 2008)[10]. The boom during the second part of this period was induced by AT liberalisation in the USA (1979) and began in Europe some years later; prospects indicate an average increase of 5% during the next 50 years. Due to the fact that, in terms of energy, AT is fully dependent on fossil fuels, it is considered as an increasing contributor to

---

[9] According to the Bureau of US Department/Research and Innovative Technology Administration (RITA) http://www.bts.gov/programs

[10] The International Civil Aviation Organization (ICAO) is the civil aviation regulating body for all matters, including Environment. http://www.icao.int

climate change and is therefore used by the authorities as a target for appropriate action. According to the report issued by the Intergovernmental Panel for Climate Change (IPPC, Brussels, 2007), the AT impact on global warming is considered one of the key issues, and probably will continue to be considered as such over the coming years.

Today, the AT industry is a global, cosmopolitan but at the same time "local minded " organisation where many central activities like those derived from ICAO regulations coexist with local or regional peculiarities and stakeholders within the subsequent conflict, both within the system and in relation to other interacting transport systems (transport modes). Since AT competition with other transport modes (in terms of trip duration) exists on distances of up to 800 miles (e. g. high speed trains), transport efficiency as a whole in usual terminology addresses the concept of "intermodality". Given the population density and distribution in Europe, it is "efficient" that the large European airports should provide intermodal facilities, as it is the case of the Thalis (Brussels-Paris Charles de Gaulle) train or the ICE (Köln-Frankfurt) and the Spanish high speed trains (AVE) that connect Madrid and Barcelona, and in the future will also connect their airports.

3. AT as a system: economical, social and environmental efficiency - a proposal of sustainability indicators

As far as sustainability efficiency of different modes of transport is concerned, widespread literature exists in which sector interests sometimes hide clear and objective conclusive results.

There is an increasing agreement to consider sustainability as a three pillar based concept of economic, social and environmental performance. In particular, the WBCSD[11] supports that the business sector should address this combined view and many companies strive to build a culture based not only on

[11] World Business Council for Sustainable Development is a coalition of 200 international companies from 35 countries and more than 25 sectors united by a commitment to sustainable development via the three pillars of economic growth, environmental protection and social equity.

economical profits but also on the achievement of societal benefits, in terms of efficient use of resources, social cohesiveness and internal democracy[12] in form of "environmental economy" or "social ecology"(Jiménez Herrero, Luis, 1996). With regard to the particular question of AT's capacity for economic survival (not the survival of individual airlines), it is assumed that in spite of the total dependence on oil and the vulnerability to global or regional crisis, it seems that AT is ready to survive, as long as the airlines continue to adapt to market requirements and the global economy is not subjected to prolonged catastrophic circumstances. However, the current world crisis is affecting the AT sector, and losses of 15% of the yearly results in 2009 will take some years to recover, particularly for small and medium sized airlines. (IATA, 2009)

If AT's overall economic capability of survival is to a large extent guaranteed, some questions derived from the environmental impact, consequence of its growth, come up: are we generating an unacceptable use of aviation? Are we willing to give up our "unjustified" habits? What is justified and what not? How can we compare "merits" of environmental burdens and react accordingly? For instance, to avoid aircraft noise in the surroundings of an airport by using a "preferential runway", an increase of 1% of fuel consumption may be necessary. Which is the priority, to save fuel (and emissions) or to reduce noise? Is it true that "no agreement can be reached as to how sustainability can be defined (and measured) in the context of the aviation industry?[13]

To achieve their mission, the organisations are provided with resources to help them reach their economic and social objectives in an efficient manner. The issue of defining the best environmental efficiency indicators for the various airline functions and the possibility of working them out with the aid of vectorial and matrix analysis has been addressed before (Sánchez Súcar, Antonio,

---

[12] The idea of social cohesiveness as a result of internal democracy has been widely expressed. Particularly in this context, in the document "Verkehr und Umwelt. Wechselwirkungen Schweiz-Europa. Nationales Forschungprogramm 41". Ernst Basler & Partners. Bern.2000.

[13] Assessment of what impacts more or less acceptable than others falls within cultural or political aspects, related to domination or influence.("Sustainability and Air Transport". University of Manchester, 2000).

2000) Now, the relevant AT subsystems have been considered in order to enable us to define sustainability indicators of several categories.

The system as a whole is depicted (Fig.2) with its inputs and outputs pertaining to the three categories: economical, societal and environmental. (Azapagic, Perdan,2000).

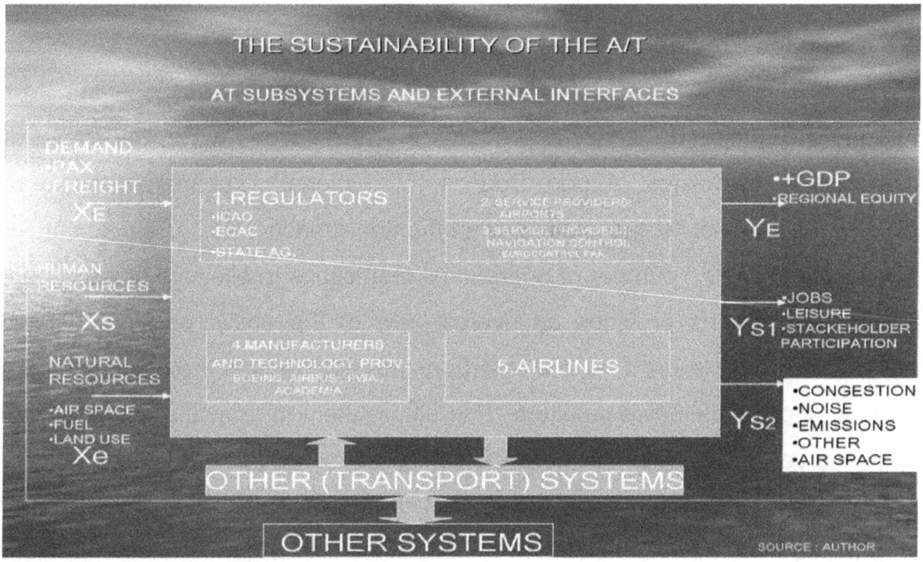

Figure 2

3.1 Regulators

The International Civil Aviation Organisation (ICAO) is the international regulating body of AT, and the Committee for Aviation Environmental Protection (CAEP) is its expertise branch. This group is responsible for the ever.increasing stringency of the environmental standards for aircraft certification compatible with the available technology. The fuel consumption per revenue passenger kilometre (RPK) of modern aircrafts has been reduced by 70 % during the last 20 years, to values around 3 liters/100RPK, similar to the consumption of a car with 2 persons on board. The reduction of the aircraft certification limits (so called Chapter 3 since 1977 and Chapter 4 since 1 Jan. 2006) reduced the acoustic footprint of modern aircrafts by 80%. The European Civil Aviation Conference, (ECAC)[14] and its group of environmental experts, Annoyance Caused by Air Transport (ANCAT), have cooperated actively with the European Commission to include the aviation in the carbon emissions trade (Directive 2008/10/CE).

3.2 Service Providers - Airports

Airports are subject to locally apply rules and regulations within the frame of ICAO/CAEP guidelines. Airport Council International[15] submit proposals to CAEP with respect to the main environmental impacts, Noise and Local Air Quality (LAQ). The number of operational restrictions to alleviate noise impact around the airports provides an overall view of the active position of local authorities on this subject. All major airports are able to map the noise contours through appropriate noise monitors. This allows them to draw the actual overall noise curves in dB with the help of mathematical models and, according to traffic growth, population census and fleet replacement, provide projections of who will be affected in future (Fig.3).

A problem arises when comparing requirements worldwide, due to the different existing metrics used[16] .The most widely used unit for measuring noise levels is the dB (A) (the A weighted scale reflects the human reaction to "loudness"), but other scales and other references exist. The perceived noise (PNdB) and effective perceived noise (EPNdB) scales incorporate the different frequencies and duration of noise patterns, resulting from various speeds and modes of operation of aircraft, the latter used by ICAO for its noise certification standards. The Maximum Sound Level (Lamax) is the maximum instantaneous value recorded of an event and the Equivalent Continuous Sound Pressure

[14] While ICAO is the international regulating body, the European Conference of Civil Aviation provides guidance at the European level. The European Commission is the ultimate regulating body in Europe. http://www.ecac-ceac.org

[15] ACI provides guidance and coordinates the airports' relationship with the regulating bodies, but there is wide discretion regarding the adoption of local policies. http://www.airports.org

[16] Boeing airport database shows the variety of metrics existing for noise measurement. http://www.boeing.com/commercial/noise

(Leq) is a measure of the average sound pressure level throughout the duration of the event. The European Community proposes "Day-evening-night level"(Lden) which is based on Leq over a whole day with a penalty of 10 db(A) for night time and an additional penalty of 5 dB for evening noise. There is no agreement even among experts on what metrics are the most representative in a particular situation. With the aim of setting a common measuring procedure, the CAEP -7 (2007) has adopted a guidance that updates the ECAC basic reference document of this issue[17].

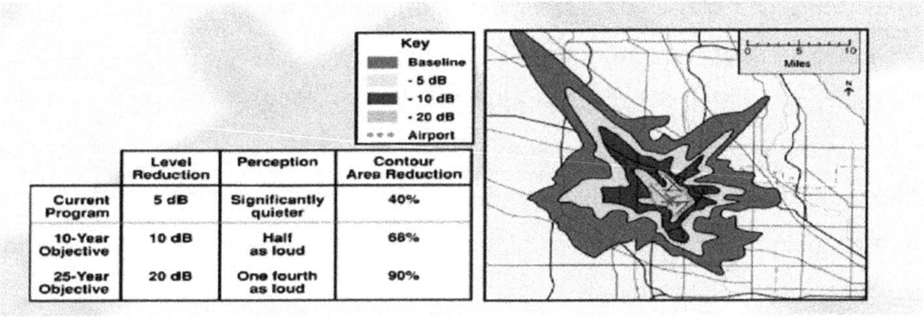

	Level Reduction	Perception	Contour Area Reduction
Current Program	5 dB	Significantly quieter	40%
10-Year Objective	10 dB	Half as loud	68%
25-Year Objective	20 dB	One fourth as loud	90%

NASA's noise reduction goals are illustrated in this graphic representation of the noise footprint around an airport.

Figure 3: Typical Airport Noise Footprint Reduction

Source: NASA

The lack of clear procedures for noise estimation in land planning affects the monetary value of assets and produces severe distortions and has given rise to a considerable amount of studies to correlate property values and aircraft noise. Researches have shown (Schipper, Youdi, 1999) that houses with relatively higher prices are subjected to higher depreciation indices due to noise exposure.

The most ambitious effort to create a worldwide model of airport noise population impact is the CAEP/MAGENTA model, carried out to assess the

[17] The document ECAC.CEAC Doc.29.Report on Standard Method of Computing Noise Contours around Civil Airports, December 2005, is the reference guideline adopted by CAEP.

benefits of the new certification standard (Chapter 4), in force since 1.1.2006, to reduce the overall limits of the previous Chapter 3 by 10 dB. This model, which takes into account projections of fleet replacement and population census around airports, predicts the affected population by Night Level DNL 55 up to the year 2020. The results show a 25% reduction of the number of people affected, in spite of the traffic increase.

In October 2002, the ACI Aircraft Noise Rating Index was adopted by the ACI´s Governing Board with the aim of:

a) Encouraging global consistency in the implementation of effective airport noise management programs.
b) Enabling airports to communicate effectively with communities and governments about noise issues
c) Providing a tool compatible with the ICAO system
d) Providing a reference point to encourage manufacturers to develop the quietest possible aircrafts.

The ACI Aircraft Noise Rating Index (Table 1) uses the corresponding noise certification data to rank aircrafts into six categories. If the aircraft meets the criteria of two different categories, the lesser category will apply.

Table 1: ACI Reference Indexes

Criteria to be met concurrently	F	E	D	C	B	A
Cumulative EPNdB reduction from ICAO Chapter 3 Standard of at least	**<0**	**0** or more	**5** or more	**10** or more	**15** or more	**20** or more
Individual EPNdB reduction from ICAO Chapter 3 Standard at each noise test point*	N.A.	**0**	**1**	**2**	**3**	**4**

* Test points are located: under approach and take off paths and at one side of the runway at take-off roll.

As far as Local Air Quality (LAQ) is concerned, there is even less overall consolidation of indexes and regulations than for those related to noise. CAEP-

7 (2007) recognises that "there is no accepted metric or modelling system to report the impact on local air quality (LAQ) emissions from aircraft, as there is for noise".

However, LAQ indexes, instead of environmental efficiency indexes, are increasingly used for environmental charges, in order to provide incentives for the use of modern aircrafts.

A matter of discussion is the contribution to LAQ from emission sources other than aircrafts, such as service vehicles in the airport platform and access traffic to the airport, that increase the total emissions depending on the configuration of airport infrastructure. An airport equipped with a sufficient number of "fingers" so as to serve almost 100 % of the flights will require a lesser number of buses and vehicles for passenger transport to aircrafts. Furthermore, the contribution attributable to ramp vehicles will be proportional to the number of those powered by electricity (which requires a complex airport infrastructure for battery loading). Road traffic induced by the airport will affect LAQ, depending on airport accessibility of public transport, in particular trains. An evaluation made at Madrid/Barajas airport before it was made accessible by city underground trains showed an additional emission impact attributed to road traffic of 15-20 % of the total aircraft emissions (Sánchez Súcar, Antonio, 1999)

3.3 Service Providers. Air Navigation Systems

The Air Navigation Service Providers (ANSP) provide air navigation services to air traffic. While the economic efficiency of such providers is determined by the service costs, their environmental efficiency depends on their capability of properly managing the demand of flights. If the system is not able to cover such demand, delays and hold-ups due to congestion of air space near the airports will occur, with the corresponding waste of fuel. In order to achieve the lowest possible environmental impact of a flight in terms of emissions, the optimal trajectory (which might not be the shortest, taking into account the atmospheric conditions, winds, and so on) must be followed, and this means less aircraft separation, which can only be achieved by means of more refined technology and the continuous innovation of procedures. To improve the efficiency of European air space, Eurocontrol[18] has launched the SESAR (Single European Sky ATM Research) project, which is to be implemented over the next few years, with the goal of providing a capacity increase of 70% in 2020 and three times more in the long term. The goal is to reach 98% of "on time" flights and <5% of flights requiring more than 2.5% fuel for contingencies - all of this at half of today´s operating cost.

3.4 Manufacturers/Technology providers

In addition to the production of aircraft, engines and general components, the manufacturers have important research and development functions to create innovation, frequently under joint ventures with official and public agencies (which is the case of Boeing or Pratt and Whitney with NASA and Airbus in partnership with some European state organisations).

As previously pointed out, the new environmental criteria put in force by the authorities must be realistic and compatible with available technology. An example of innovation is the project to develop combustor technologies carried out by NASA Glenn Research Centre [19] and Pratt and Whitney, in order to reduce landing and take-off emissions of NOx by 45% in 2016 and by 70% in 2026, and the NASA Quiet Aircraft Technology to curb perceived aircraft noise by 50% in 10 years and by 75% after 25 years. The predominant use of lightweight composite materials in building aircraft structures promises significant weight reductions, with the subsequent fuel and noise benefits. In the new Boeing 787, fifty percent of the primary structure - including the fuselage and wings - will be made of composite material, potentially reducing fuel consumption by 20%.

The new generation biofuels are the most promising alternative to oil fuels, because of their smaller carbon footprint. Unlike first generation biofuels,

---

[18] Eurocontrol is the Air Navigation Service provider in Europe. http://www.eurocontrol.int
[19] See http://www.nasa.gov

modern biofuels are made from sustainable non-food biomass sources such as algae, switch grass and other plants and do not compete with fresh water requirements or food production. The potential of hydrogen as an alternative fuel is being studied, but due to the major changes in aircraft design required and the challenges related to security and logistics, it is expected to be operational only in the long term.

3.5 The Airlines

The typical environmental impacts from airlines are noise at the airports and in their vicinity, and emissions, both at airport level (LTO)[20] and cruise altitude.

Since noise and LTO emissions have local effects, the corresponding indicators have been attributed to "Airports" and only those produced during cruise are attributed to the "Airlines". Cruise emissions are directly related to fuel consumption, one of the main costs drivers on this subsystem, and are an important contributor to climate change. The report "Aviation and Global Atmosphere" first issued in 1999 by the Intergovernmental Panel of Climate Change (UNEP/IPCC)[21] predicts an increase of the radiative force attributed to AT (i. e., the energy balance of the Earth-Atmosphere system in watts/m2) by a factor of 4 in the year 2050, compared to the 1992 values.

An important way for airlines to improve their fuel efficiency and reduce emissions is to maintain high load factors, optimising their network, fleet mix, aircraft configuration and frequencies. Fuel efficiencies and emissions reductions at cruise altitude can also be achieved through aircraft maintenance and operations, for example, reducing aircraft drag by cleaning the airframes and engines, minimising the number of non-revenue flights and flying at the most fuel-efficient speeds and altitudes.

Four major factors restrict operational decisions or procedures: safety,

---

[20] The Landing and Take Off (LTO) Cycle comprises operations in and around an airport (taxiing, take off, climb, approach and landing) up to 900 m altitude.

[21] The reference report of cruise emissions impact on climate change (while LTO emissions affect LAQ in the vicinity of the airports).

legal constraints, environmental trade-offs and specific situations. Operational opportunities can only be considered in the context of airport and air traffic constraints.

4. Proposed Indicators

Table 2 shows a set of some potential AT indicators relevant to "Nature Conservation", "Freedom of Mobility" and "Material Wealth" under the frame of Reference Pattern of Values[22] in categories of MPI (Management Performance Indicators) and OPI (Operational Performance Indicators), directly related to production (ISO 14031). As we pointed out, efficiency can be calculated as the output /input ratio or the ratio of two outputs represented as Y_E and Y_X. In this case "Material Wealth" is a function of an input (i.e. Aircraft Movements or "Stock Capital" is a direct function of the global input "Demand").

The indicator ACMOV/ Km2 for the 65 dB(A) noise curve (as the "day" reference value generally accepted) refers to the number of aircraft movements (ACMOV) causing this noise reference footprint. This indicator accounts for the footprint at this sound level, as a direct indicator of AT performance which cannot be clearly related to the indexes of affected population (land planning around the airports falls under the responsibility of the local or regional authorities since AT does not have complete control over it).

As far as Local Air Quality (LAQ) is concerned, states have historically developed their own air quality guidelines, although some airports apply more stringent criteria. The emissions normally addressed are nitrogen oxide (NOx), hydrocarbons (HC), carbon monoxide (CO) and sulphur oxides (SOx), with NOx being considered as the most relevant of aircraft emissions at airport level. The emission levels per LTO and cruise can be measured using the ICAO emission databank (that relates fuel burned to the different pollutants, different operating modes, take-off, taxiing, etc.). For nitrogen oxides, the index of a given fleet

[22] The "Reference Pattern of Values" comprises a set of universal values: Health, Security, Material Wealth, Knowledge, Freedom (of Mobility), Justice, Nature Conservation, Quality of Activities and Prestige. (Max-Neff, Parra Luna 1993)

would be the aggregation for each operating LTO mode is given as:

$$EI_i = NE_i{}^* \sum \quad T^*FF^*EI_{NOx\,i}$$

Table 2: Proposed Sustainability Indicators (at subsystems)

Subsystem	Typical Input	Members	Typical Output	Indicator	Type	Output Class	Remarks
1) Regulating	1.1A HR (N1i)	ICAO, ECAC, States, Agencies	1.1B NPRO 1	$\sum \Delta p1i/p1i^* Ki^*1/N1i$	MPI	Yx	
2) Service Providers	2.1A HR (N2i) **1.1B** **3.1B** **4.3B**	Airports	2.1B NPRO2 2.2B ACMOV 2.3B NOISE 2.4B LAQ	$\sum \Delta p2i/p2i^* Ki^*1/N2i$ ACMOV/N2 ACMOV/KM2 $1/ACMOV^* \sum NRli^*ACMOVi$ $1/ACMOV^* \sum ACMOVi/EIi$	MPI OPI OPI OPI OPI	Yx,Y_E Y_E Yx Yx Yx	LAQ: Local Air Quality
3) Service Providers	3.1A HR (N3i) **1.1B** **4.3B**	Navigation Control Centers (FAA, Eurocontrol, etc)	3.1B NPRO3 3.2B MF 3.3B DF 3.4B HLD	$\sum \Delta p3i/p3i^* Ki^*1/N3i$ MF/N3i DF/MF HLD/MF	MPI OPI OPI OPI	Yx,Y_E Y_E Yx Yx/ Y_E	
4) Manufacturers and Technology Providers	4.1A Stock Capital 4.2A HR (N4i) **1.1B**	Airbus, Boeing, PWA, CFM, NASA, Academia	4.1B Aircrafts (AC) 4.2B Engines (ENG) 4.3B Technology	$\sum$ ACi /(SCUi*N4i) $\sum$ ENGi/(SCUi*N4i) $\sum \Delta p4i/p4i^* K^*1/N4i$	OPI OPI MPI	Y_E Y_E Yx,Y_E	
5) Airlines	5.1A Stock Capital 5.2A HR (N5i) 5.3A AC **1.1B** **2.1B** **3.1B** **4.3B**	All IATA and not IATA Airlines	5.1B ATK 5.2B NOISE 5.3B CREM	$\sum$ ATKi/(N5i* SCUi) $1/ACMOVA^* \sum ACMOVAi^*NRli$ $1/ACMOVA^* \sum ACMOVAi/EIi$ $\sum$ (FU)i/RTKi	OPI OPI OPI	Y_E Yx Yx	CREM= Cruise (non LTO) Emissions

HR = Human Resources

AC = Aircraft

ACi = Aircrafts of fleet i

NEi = Number of engines of ACi

NPRO = Procedures,guidelines issued

$\sum \Delta p/p$ = % of improvement of an environmental impact (noise or emissions reduction) following a procedure

K = Weighting factor, for local (1), regional (10) and global (100) effect

Ni = Number of people involved in a project (AC type, task force,etc)

ACMOVi = (Aircraft Movements per AC Type)The sum of the number of take offs and landings of AC´s with same NRI or $EI_{NOX}i$ in an airport

ACMOVAi = Idem of a specific airline

ACMOV = The total number of Movements of an airport

ACMOVA = Idem of a specific airline

NRIi = Noise rating index of Aircraft **i.** 1 for Category F increasing to 6 for category A

$EI_{NOX}i$ = Emission Index of ACi for each LTO Mode (take off, landing, taxiing, climb and approach) and cruise in gr. of pollutants per KG of fuel (As reference the Nox index . Indexes of other emissions to be used)

EIi = Emision index of ACi = $NEi^* \sum_{\text{LTO MODE}} T^*FF^* EI_{NOX}i$

EEI = Emission Efficiency Index=1/ACMOV*∑ACMOVi/EIi

FF = Fuel Flow in Kg/min used by ACi at each LTO mode.

T = Time in minutes of the AC operation during each LTO mode

F = Managed Flights

DF = Delayed Flights

HLD = Flights requiring Holding before final approach

SCUi = Stock Capital Unit of project **i** (or fleet type)

ATKi = Available Tonnes Kilometre per fleet. This is the main parameter used to measure the productivity of an airline

RTKi = Revenue Tonnes Kilometre per fleet. Takes into account the "load factor" of the fleet **i.** (LF is RTK/ATK=1 if full capacity is used)

FU = Fuel used by fleet in cruise. This is the total fuel used published in the yearly Environmental Reports of the airlines minus the LTO fuel

KM2 = Square Kilometre of 55 dB footprint

Black characters mean inputs from other subsystems outputs

would be the aggregation for each operating LTO mode is given as:

$$EI_i = NE_i^* \sum_{\text{LTO MODE}} T^*FF^*EI_{NOx\,i}$$

And the global emission index takes into account the number of movements $ACMOV_i$ of each fleet and the overall AC movements at the airport

Emission Efficiency Index= $1/ACMOV^* \sum ACMOV_i/EI_i$

For manufacturers and technology providers, the production indicators in pure terms (of Y_E form) are regularly used in the Financial Reports, while the relevant indicator for airlines is the fuel consumption at cruise per revenue ton kilometre (Ys / Y_E). Fuel consumption at cruise is the total fuel consumption minus LTO consumption.

5. Information and Transparency

The concern of global society regarding ecological disasters (Beck, Ulrich, 2002) and the increasing demand for ethical values have prompted the companies of the business sector to seek new sensible segment markets with the corresponding orientation towards a culture of corporate responsibility and green image as a powerful competing factor.

One might think that "image" or "prestige" related issues are "virtual products" belonging to the realm of perception and that a preliminary question should be formulated: is it possible to build corporate prestige while ignoring the demand for a truly socially responsible and measurable performance? What are the means by which the organisations report and make credible such behaviour? The literature and all involved actors agree that the key element of such an issue is Sustainability Reports or Corporate Responsibility Surveys (CRS). These are the documents where the policy, programmes and corporate goals are shown and audited by rating organisations using objective criteria to create a new market where business companies, ranked by their excellence in responsible performance, are evaluated by investors and other stakeholders[23]. However, nowadays, with more stringent criteria for incoming companies having been established over the last few years, because of the increasingly responsible performance of the top leaders, there is considerable room for improvement.

A survey carried out by one of the largest business research centres in Spain (ESADE, 2007), which included data from ten countries, showed that responsible investment accounted for 7 billion Euros in the UK (year 2005), 3 billion in France and 1 in Spain, reflecting the fact that 84% of Spanish investors are not willing to invest in non-responsible companies.

The largest worldwide effort to promote and provide guidelines in this direction is being undertaken by the Global Reporting Initiative (GRI)[24], a UN-sponsored programme that provides guidance for assessing economic, social and environmental performance through a common framework intended to compare performance among organisations. The following reflects the CRS published under GRI principles since 1999:

1999	2000	2001	2002	2003	2004	2005	2006	2007	2008
11	46	124	141	176	288	377	517	691	1044

However, a survey of the Corporate Register Data Bank[25] shows that, although the number of published reports has increased worldwide from 100 in 1993 to more than 3000 in 2008 (333 of those pertaining to the 500 largest companies), only one third follow the criteria given by GRI. It must also be pointed out that only 11% of CSRs from the surveyed companies showed that a consultant had checked the report. From these figures it can be seen that, in spite of the accelerated progress on this direction, there is a general lack of common

[23] There is a "market" of Social Responsibility Index providers , like the Dow Jones Sustainability index http://www.sustainability-indexes.com., FTSE4Good. http://www.ftse.com/ftse4good or Business for Social Responsibility. http://www.bsr.org

[24] Global Reporting Initiative provides guidance for criteria to be used on Sustainability Reports. http://www.globalreporting.org

[25] The Corporate Register is the World´s largest directory of sustainability reports. http://www.corporateregister.com

standards and transparency in the assessment of sustainability performance.

Concerning AT, the available data indicate that the practice of environmental reporting is almost exclusively restricted to the top 20 IATA member airlines as classified by revenue ton kilometres (RTK)[26]. These figures were confirmed in 2007 through the individual airline web pages of the 251 IATA member airlines, only 26 of which have made the reports available through hard copy or via the web.

Conclusion

The first question in connection with the title of this paper concerns the concept of sustainability itself: "processes are sustainable when the deterioration rate that they produce in the environment is lower than its recovery rate". This definition leads us to accept that AT, like all other modes of transport, as well as all other industrial activities involving combustion processes, is environmentally unsustainable. A more detailed look at the problem should include questions of how fast deterioration occurs, what the tendencies are and what means of measurement and control exist to diminish and eventually reverse the burden. One could pose questions such as: up to what extent should the right of mobility be preserved in view of the undesirable by-products affecting nature conservation? How important is the consumption of products that must be transported from remote regions if this contributes to these regions' economic development in an ideal worldwide trade (material wealth), in view of the emissions produced during their transport (nature conservation)? What about the need of access to remote regions in developing countries, where surface transport is scarce or non-existent?

Both historical data and future projections indicate that although AT, as a closed system, is not sustainable in absolute terms, its tendency towards relative sustainability (or efficiency) is clearly positive. That is to say, the emissions or the noise footprint per unit of societal product (mobility measured

[26] IATA Airline Environmental Reporting Survey, 2001 http://www.iata.com

as RPK) are constantly decreasing.

As a further step, a compensation mechanism envisaged by the Kyoto protocol, "carbon emission trading", means that activities dependent on fossil fuels like AT will be allowed to "buy pollution" from sector able to implement new green technologies and infrastructure innovation. The new EU Directive 2008/101/CE incorporates European AT in the carbon trade regime recently extended to aviation. According to IATA,,this, ,among other technological and operational factors, will permit aviation to cap CO2 emissions by 2020 (Fig.6)

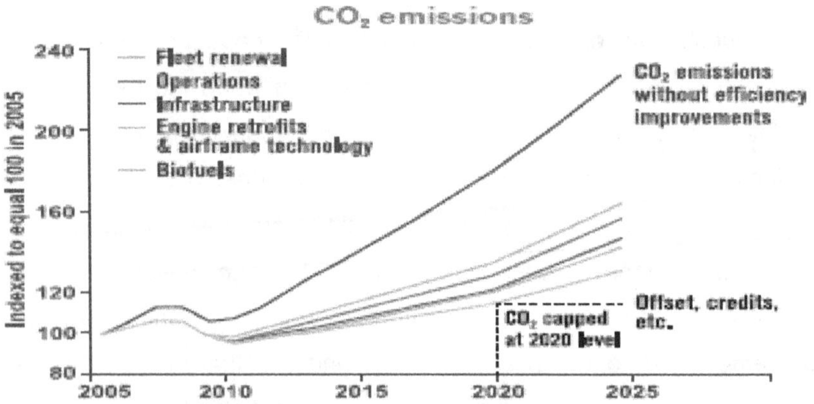

Figure 6

In view of the above, it can be said that:

a) There is a sufficient set of AT sustainability indicators, in particular those specifically related to the main environmental impacts (climate change, noise) concerning which historical trends show an ever increasing efficiency. However, those affecting local conditions, like community noise and local air quality, are, not surprisingly, subject to local decisions that, in order to prevent unbalanced policies among airports, should be properly coordinated.

b) In order to make its goals effective and workable, AT should be considered as an open system. Efforts to reduce community noise will be offset if measures to avoid airport encroachment are not put in force by land planning authorities. Similarly, local air quality standards for aviation will be ineffective without providing efficient ground transportation systems to access airports, which are far beyond AT responsibility.

c) Efficiency (impact per transportation unit) is not sufficient to justify the sustainability of AT, and absolute limits must be set up with regard to local air quality, climate change and noise.

d) Internalisation of external costs will have to be undertaken with regard to all activities that have an environmental impact. In the case of climate change, in conjunction with the principle c), the aim to include AT within the carbon trading scheme seems to be appropriate. Such an internalisation should be done at the lowest possible cost to permit emission rights exchange with sector participants not entirely dependent on fossil fuels.

e) As far as product information and transparency are concerned, CSRs have gained a relevant place as outstanding tools for the enforcement of more ambitious sustainable goals, but there is still room for improvement. There is an increasing tendency to adopt common standards, not only through the GRI Guidelines, but further progress has to be made to ensure that reports are produced on a comparable footing. Often it is difficult to compare different reports from different sized companies and varying organisational cultures. Environmental reporting has not yet reached the same degree of public acceptance and credibility as financial reporting, possibly because of its more generalised presentation of environmental issues and related risks, rather than indications of performance levels.

f) Last but not least, the question is: should demand be accommodated? It seems that the answer is yes, whenever other transportation or more efficient alternatives are available. Intermodality is one key question that infrastructure planners must consider. As an example of this, the European network of high speed trains will foster alternative options, alleviating the congested European air space to reserve it for use of long distance flights, or others, where no alternative to the airplane exists.

Finally, one can say that the future of air transport sustainability will largely be affected by the two relevant forces affecting AT: technology and growing demand. On the one hand, let us salute the IATA goal to adopt a CO2 emission cap by 2020 with a reduction of 50% by 2050, based on 2005 levels. The answer of Michael Levine, a researcher of New York University, sums up the problem: "With the prospects of increasing air traffic, between the aging population of the OECD countries, the emerging markets and the continuing globalisation of business activity, this can make us sweat. From a technological point of view, I don´t believe in zero emissions. But I believe we have to be responsible for our emissions and declare to the public that we are going to make the best efforts we can". The controversy exists (and will continue to do so as long as our current dependencies remain).

References

Andersen, M.M., Tukker, A., (ed.) (2006) *Perspectives on Radical Changes to Sustainable Consumption and Production SCP*. In 'Proceedings of the Workshop of the sustainable consumption Research exchange SCORE! Network'. 20-21 April, Kopenhagen Dänemark.

Azapagic, A. Perdan, S. (2000) "Indicators of sustainable development for industry: A general framework". Process Safety and Environmental Protection 78, 243-261.

Beck, Ulrich (2002) *La Sociedad del Riesgo Global.* Siglo XXI Editores.

Daly H.E. (1991) *Steady State Economics.* Washington D.C.: Island Press.

ESADE (2007) *Report on Sustainable Investment.*

Grimley, Paul M. (2006) *Indicators of Sustainability Development in Civil Aviation.* Loughborough University.

Janic, Milan. (2007) *The Sustainability of Air Transportation. A quantitative Analysis and Assessment.* Ashgate.

Jiménez Herrero, L. (1996) *Desarrollo Sostenible y Economía Ecológica. Madrid:* Editorial Síntesis.

Oxford Economics for ATAG (2007) *The social Benefits of Aviation.*

Parra Luna F. (1982) *Elementos para una Teoría Formal del Sistema Social.* Universidad Complutense de Madrid.

Sanchez Sucar, A. (1999) *Las Emisiones no Aeronáuticas en los Aeropuertos.* III Congreso Nacional de Medio Ambiente. Madrid.

Sanchez Sucar, A. (2000) *Environmental Balance and the Performance of Social Systems.* Kluwer Academic/ Plenum Publishers.

Schipper Y. and Rietveld P. (1999) "Why do aircraft noise value estimates differ? A metaanalysis". *Journal of Air Transport Management* 4, 117-124.

SDC (2001) *Report on Aviation Sustainability. London: Sustainable Development Commission.*

SDC (2006) *Redefining Progress. Report of the SD Panel Consultation on Progress.* London: Sustainable Development Commission.

United Nations (2007) *Indicators of Sustainable Development Framework and Methodologies.* United Nations Publication Sales No. E.08.II.A.2, New York.

Upham, Paul, et al. (2003) *Towards sustainable aviation.* James & James/Earthscan.

7

Efficiency! To what end?: Considerations on how to increase the chances of survival of the human race

By Elohim Jiménez-López

Abstract

After exploring, in current literature, the question of how the concept of efficiency emerged as an argument, I offer a concise examination of three main concerns:

1) How did the *human species fail to evolve efficiently?* Its failure to do so is mainly due to two creeds: "some humans are the *owners* of the planet"; "the majority of humans must *perform homogeneously*".

2) The *inefficiency* of the means – in time and in space – employed by humans for *survival;*

3) The crucial question: how can we *invent* new *efficient* ways and measures for *humanising, in a responsible manner, the dehumanised homosphere.* I also offer a brief summary of the methodological research that is required. I conclude with a few selected quotes from Mahatma Gandhi.

1. Introduction

Something effective, effectual, efficacious, efficient... is generally defined as something that produces a desired effect. The need to improve industrial processes in order to ensure the development of productive forces conceived and implemented for the purpose of satisfying more and more of the requirements and concerns attributed to people seems to have given rise to the assumption that "...it is necessary to measure industrial efficiency". This assumption became an essential factor in all decisions made in the course of the quest to improve living standards during the second half of the twentieth century. But civilisation as a whole has become the source of greater and more numerous difficulties that have been disrupting the homosphere, the biosphere and the ecosphere for a number of decades now. Therefore, since industrialisation is an essential factor of civilising dynamics, it is necessary to define the related concepts: effect, cause, effective, effectual, efficacious, efficient, efficacy, efficiency.

Effect is a result or condition produced by a "cause" or "something that happens when one thing acts on another" (Fowler,1926)

Cause is "something which produces an effect" or "an event that makes something happen" (Longman, 1978)

Effective is the event "capable of producing a result" or something "producing the desired result" (Longman, 1978)

Effectual (of an action) "producing the intended effect" (Longman, 1978)

Efficacious of "a course of action...producing the desired effect in dealing with a problem" (Longman,1978)

To be *efficient* means to work well, quickly and without waste (Longman, 1978). It would improve our efficiency if we used more up-to-date methods (Longman, 1978)

Efficacy is the power of producing an effect (Merriam-Webster, 1978)

Efficiency is the capacity to produce desired results with a minimum expenditure of energy, time and resources (Merriam-Webster, 1978)

Efficiency is the ratio of the work done or energy produced by an organisation or machine to the energy supplied in the form of food or fuel, ... brings one's ability to bear promptly on the thing to be done, ... competent power. Power is the right ability to exercise control, legal authority, capacity (Funk&Wagnalls, 1970)

Efficient is the term used to describe a person who duly accomplishes the function assigned to him/her (Moliner, 1984).

But none of the eight words described (effect, cause, effective, effectual, efficacious, efficient, efficacy, efficiency) were included in Samuel Johnson's Dictionary (1775); it may therefore be assumed that its eventual inclusion reflects the increased expansion of industrial work since the end of the XIX century.

However, although very often, "efficiency is exclusively related to economic affairs", we can still argue that "a machine is more efficient if it generates more power or more products for a given amount of fuel or input". Logically, "economic efficiency ... addresses itself broadly to the goal of bringing the limited resources of society into proper relation with the desired ends. The ends themselves are expressed through the patterns of social valuation or demand for finished goods, which pattern is determined in part by the market mechanisms of the private sector of the economy and in part by the political mechanisms that control expenditures in the public sector." (Kuper & Kuper, 1985)

Besides, the requirement for efficiency is justified by the argument that it is "... needed... people do not want to waste scarce resources ... they have priorities among various things they are trying to achieve." "The main purpose is the human and just treatment of people ... who are having rather a hard time anyway."

It seems proper, therefore, to introduce a new way of doing things that appears likely to work at least as well as the old way and is at the same time less costly (setting free scarce resources that can be used to provide other good things.)

"...the antipathy to 'efficiency' stems from much narrower interpretations: ...there is a brand of 'efficiency-mongering' which is obsessed with saving money (as if that were the only scarce resource), irrespective of the effects on performance".

At the other end of the scale, we find equally blinkered individuals who think only in terms of performance and pay no regard to cost. "...it is not necessary to be 'hard-headed' in a commercial sense ... but it is necessary to be 'hard-thinking' in a humanitarian sense, assuming the objectives of services, how we know they are being fulfilled and what weight we give to one objective compared with another". Additionally, there are proposals to reject simple market mechanisms such as financial profitability for determining what the priorities should be" (Williams & Anderson, 1975).

Anyhow, efficiency as a conceptual argument is an intellectual tool – invented recently - for determining how the performance of industrial tasks could be improved in accordance with certain chosen criteria. Yet this is the outcome of an idealised reasoning that prevents recognition of the fact that the whole trajectory of civilising concerns – during the last millennia – has been a fallacious course due to reasoning aimed at the justification of arbitrary (more efficient) results from actions organised by increasingly selfish decision-makers. In fact, this reasoning has determined the features of civilisation dynamics that prevail until today, under the assumption that this is the perspective available for the progress of humanity.

Consequently, these dynamics have been the encouraging force of the massive industrialisation that has engendered the belief that everything on Earth may be regarded as resources to be transformed or exploited for the satisfaction of the needs of educated humans who are willing to be increasingly modernised. At the same time, "masses of humans" as living entities must learn and relearn compulsively how to be forever efficient buyers and consumers of all that can be produced for the sake of business by disciplined and obedient workers.

Efficiency is considered a particular methodical approach aimed at improving repetitive actions in order to push ahead the civilising process. However, if we examine the kind of improvements aimed at reaching "industrial efficiency" that were increasingly implemented in the 1950s, it becomes obvious immediately that these were motivated by financial concerns both in theory and in practice.

"Whatever the disadvantages of a system highly competitive between a large number of small independent units, it can be claimed that the grossly inefficient units were eliminated in the competitive struggle. As soon as the element of competition between small units is eliminated, some other basis for judging effectiveness must be found." "... business efficiency as a whole is a combination of efficiencies in various spheres; ... There must be a combination of designing efficiency, acquiring efficiency, production efficiency, selling efficiency and financial efficiency. ... co-ordinated by organising capacity and administrative ability backed by capable, willing and co-operative workers".

"As a general measurement of the overall success of a firm in a competitive market, no adequate substitute seems to have been found for 'net profit' and its relation to various elements" "... in a competitive world, a good past record of reasonable net profits was substantial evidence of probable efficiency." Eventually, it is claimed, the efficient manifestation of industrial activities will also improve other things for society as a whole.

"There is a growing tendency to regard any established business (which has been built upon the enterprise of the owners, the skill of the employees and the demand of the consumers) as a social entity probably necessary to the community." "An increase in our productive assets should, in the long run, increase the national standard of living, for they create the potential of an increased output of goods. If the amounts taken out of profits by way of taxation are used for the creation of productive assets, e.g. hydro-electric schemes, etc., this will also result in an increased supply of goods and services; similarly, if these amounts are used to build new schools or hospitals, there may be a permanent increase in amenities, and, presumably, a rise in the standard of living." (Scott, 1950)

Purposeful efficiency is conceptually clear; however, the effects that it generates day after day are increasingly controversial because the dynamics of industrialisation have been efficiently commercialised, while the many societal incongruities resulting from it are systematically ignored. Throughout the millennia, humans have become increasingly aware of the possibilities offered by their natural surroundings while learning to make more and more tools that enabled them to give free reign to their inherent opportunism and utilitarianism. The development of this attitude soon led some minds to believe that certain humans would live "better" by taking unilateral advantage of everything found in their natural surroundings, as though everything were a resource waiting to satisfy human ambitions.

This utilitarianism, encouraged by the magnificence of the natural environment, led many to assume that there will always be plenty of resources to satisfy human wishes. So far so bad - because this attitude has motivated those selfish individuals who managed to achieve positions of power in one or another particular community to consider most of their fellow men as "human resources", necessary for transforming natural resources into facilities for empowered minorities.

At present, most societies - governed by humans of high rank - consist of groups of lower classes of humans who are systematically "trained and educated" by compulsory means to become slaves, servants, vassals, workers, soldiers ... industrialisation consolidated this attitude, and also caused human existence to become unsteadily dependent on nature on this planet because the diligent industrialists refuse to notice that within the universe, the earth is a unique BUT restricted space for the living members of many species, not only the human one.

All members of every living species that has emerged due to the action of evolutionary forces within this small space must learn to collaborate and to compete, though unconsciously, for survival supported by limited natural resources. Of them all, only humans are able look to look ahead and realise that their own future may be at risk when the regional factors of the homosphere are manipulated so as to justify the increasing use of non-renewable resources that - "God willing" - will never be exhausted.

2. The "civilised" advancement of mankind has been systematically disrupted

In the course of only a few millennia, several tribes of our barbarous ancestors, - those who had been able, on their own, to develop human features - managed to settle down here and there, apparently as a result of the successful invention of agriculture, as a culmination of collaborative tasks.

It seems obvious that the minds of most of our ancestors evolved while each one – individually or collectively through collaboration - managed to learn different ways of perceiving and reacting to diverse aspects of their immediate surroundings. Apparently only 30 millennia were necessary for the emergence of Homo sapiens sapiens - his presence is confirmed by the evidence of the nomadic wanderings of individuals. The behaviour and actions of these individuals, though, must have been restricted to the effort of surviving some few days ahead, because:

1) they were unaware of the consequences of their own actions;
2) they did not care what these consequences would be;
3) they cannot have had any idea that their behaviour would genetically and culturally influence the ways new generations might develop in the future.

Wandering helped every individual to recognise and organise his/her behaviour in the face of many different circumstances.

Meanwhile, the universal evolutionary forces on our planet were at work, determining the features of all other living beings competing for the scarce resources. However, most humans failed to notice the evolutionary forces that had determined their development. As soon as a group of our ancestors managed to settle down here or there, there were some individuals among them who subscribed to two fallacious creeds: *ownership* and *homogenisation;* this development took place simultaneously in different locations.

2.1 Ownership

It is not known precisely how the idea of ownership was conceived; but it emerged inside settled communities over several millennia. It can be assumed that, at some point, a member of a group - supported and encouraged by a few others - conceived first that he could and later that he should be the *owner* of some of the objects in his surroundings: natural resources and man-made things.

Additionally, other humans (men and women, children and adults), who were also members of the group, came to be considered objects belonging to the designated proprietor, by all kinds of reactionary means for being unilaterally integrated. Today, we have plenty of historical evidence of the way humanity was forcibly turned into "civilised entities".

The persons who developed the concept of ownership were necessarily members of an elite group, who pretty soon learned efficiency in acquiring privileges as the owners of increasing quantities of objects, people and land.

Did the ways these pretended rights were invented by particular minds unavoidably encourage or justify the process of settling people here or there? Or did the emergence of every settlement, though each one was peculiarly conceived, organised and implemented, inevitably generate the various ownership domains that characterise each civilising experience?

Whatever may be the right answer to the question of what happened in every location, these events that repeated themselves throughout the historical trajectory of humankind always supported the status of the empowered elites, who considered them as the reasons that should determine the dynamics of the various civilising processes.

Archaeological finds and written statements, supported by religious interpretations, have (in recent years) led some of our intellectually minded fellow men to infer that almost every proprietor has always been fascinated by his or her possessions.

2.2 Most humans seem to be inherently conservative

Today it is clearly evident how religious ideas were introduced, maintained and reinforced successfully in every society, in order to help people accept the destiny mapped out for second and third order people by the elites.

The whole civilising process was due to human conservatism, the diversity of which has been the source of many different kinds of fundamentalism. However, the increasing abusiveness generated by some members of one or another elite – persons who were a lot more selfish and greedy while attempting to get more efficiently civilised facilities - has caused and continues to cause the emergence of rebels, insurgents, revolutionaries, secessionists, dissenters, heretics... dissidents who dared to choose, despite the risks, to fight for the recognition of their right to ethical development of their individuality.

Over the last few centuries, the scientific and industrial revolutions have required the support of philosophical and religious arguments for justifying ownership. Until recently, their reasoning was aimed at encouraging the masses to learn and assimilate fabricated appraisals of civilising dynamics and what should be the role of humans in every society. At present, many scientists, researchers, scholars and other intellectuals are paid salaries and allotted research funds provided by those who are in powerful positions in the so-called developed countries. They are empowered decision-makers that search systematically how to continue taking unilateral advantage of natural and human resources to their own advantage.

2.3 Civilised homogenisation

This has been the policy of empowered groups ruling developed countries. It has prevented the improvement of diverse natural events, things and living beings that emerged and exist in time and in space due to the mysterious ways the evolutionary forces have been acting on Earth. It is a fallacious invention.

Only one sun has helped the evolutionary forces to manifest life on Earth since around 4.500 million years ago. Around 3.500 million years ago, the light and heat of this burning star and something else mysteriously created the conditions for the emergence of life – through self-organising entities that could reproduce themselves on their own. During the last 2.000 million years, vegetation has proliferated everywhere and created the environmental features of this planet that provide uniquely suitable conditions for the presence in time and in space of many diverse living species.

And it just so happened that some ancient hominids managed to develop what became human features. This occurred without any action on the part of other living "things" that inhabited particular locations of the Earth. They were creatures who already enjoyed being alive, though they suffered from time to time when they were obliged to search for and create circumstances on Earth for organising their survival. This development took several million years, and there was no way of predicting whether humans might emerge efficiently or not, efficaciously or not from their natural circumstances.

Millions of humans should be encouraged to learn how to recognise the uniqueness of this Earth moving around this Sun, which some humans believe is a single burning star located somewhere on the Milky Way. Unfortunately, until today, the vast majority of humans does not care about the presence of this body in the firmament, except when they complain about the "dog days" in summer time lasting longer and longer each year and attribute this to "climate change".

Anyhow, these are undeniable facts, though every human mind perceives them in a different way:

1) because the sun's light and heat that reaches every earthly mountain, coast, desert or jungle ... is conditioned peculiarly by particular atmospheric circumstances that interact in accordance with many diverse dynamic trends. They become the causality that determines the magnificent diversity of the terrestrial landscape.

2) because every human perceives sunlight and heat without knowing what other humans perceive. A genetic predisposition, inherited peculiarly by every individual, helps us to learn how to develop or hinder personal perceptions; to enjoy sunlight and heat or to suffer it with resignation. Everyone is unavoidably trained and/or accustomed by their own physiological possibilities and psychological experiences;

3) because every human, while learning to live better, searches opportunistically how to survive.

These three questions trigger different reactions in every single mind, because none of us can explain what could be perceived by other minds. Anyway, there are not two trees, flowers, fruits, animals, rocks, waters, types of soil ... that are completely alike:

a) because every living organism is the outcome of unique genetic features that have determined the particular evolution of previous organisms, those that could survive successfully;

b) because the evolution of every living being may be unexpectedly affected, disrupted or stimulated, broken or encouraged by external forces from its particular environmental conditions.

c) because the integration of soil, waters, minerals, rock ...is, in each case, the outcome of a unique phenomenon.

For these reasons, every human being as a person is a unique individual. So far so bad, because the dynamics of the whole civilising adventure have been determined by unilateral utilitarian actions most of which have been aimed at rendering the performance of most humans artificially alike, though preserving a clear distinction between:

a) the functional and efficient use of masses of people as employees, sometimes to exploit them as a labor force and sometimes to use them as reliable buyers and consumers, needed for the accomplishment of so-called civilising concerns. They constitute the potential workforce needed for sustaining the dehumanised trajectory of all civilising attempts. The functionality of this force is determined by a sufficient number of slaves, soldiers, workers, employees ... identically "educated" and uniformly "trained" for maintaining, increasing and improving the facilities to assure the total and permanent enjoyment of the members of the privileged classes;

b) the functional enjoyment of the members of the privileged classes who must buy the luxurious merchandise.

The efficiency and reliability of the ways in which we are being civilised have certainly been determined, dogmatically but effectively, through the training and schooling of individuals, the marketing of goods and services, the culture of communities, the evolvement of economic and sociological concerns, the governance of communities, urban development, the performance of nations... Civilised decision-making has succeeded in making everything in the homosphere - including humans - artificially homogeneous; therefore most people pretend to ignore that natural circumstances are always inevitably heterogeneous. Numerous archaeological findings - confirmed by historical information – show how empowered minorities have continuously organised utilitarian actions, disregarding the diversity of nature and humankind, while aiming at a purposeful organisation of the homogenisation of natural circumstances and people. For several centuries, up to the present day, everything has been considered homogeneous for the sake of continuously increasing financial profits through efficient marketing and financial speculation of products and services invented and produced solely for attaining such profit.

A few millennia of civilising attempts have disregarded the long stretch of time that our predecessors needed to develop their unique features. Instead, decision-makers invented and imposed "diverse" methodological procedures, such as religious, economic, political, governmental, nationalistic ... principles, doctrines, fundaments, dogmas ... aiming "diversely" at motivating, suggesting, convincing ... obliging masses of human brains to "learn" that the progress of any "civilised" society requires indispensably homogeneous thinking, behaviour and performance. In addition, creativity is encouraged in order to invent a modernised racist ideology, justify a new nationalism, propose another social class... The question what humans are or should be is answered through fallacious interpretations.

The future of our "magnificent" civilisation is arbitrarily announced through the manipulation of human minds, using the "developed" scientific discipline of behavioural engineering for preventing millions of minds from thinking on their own. Paradoxically, this serious problem is generated by the dynamics of modernised civilisation.

The manipulation of human minds is even encouraged through the scientific "development" of behavioural engineering (Bertalanffy, 1958), and purposefully conceived and employed for preventing millions of minds from thinking on their own. In the 1960s, Ludwig von Bertalanffy frankly and clearly denounced the danger generated by this perverse engineering aimed at increasing the control over possible ways of exploiting human resources. This manipulation intended and still intends to decerebralise humans for the sake of a better control over the ways human resources can be exploited. So far, Bertalanffy's warning against robotomorphism and zoomorphism which are aimed at an efficient and successful training of humans, in order to make them behave as conditioned-response robots, while claiming that these second-order humans are incorrigible naked apes, has practically not yet been taken into account.

The intention of decerebralising humans stems from the assumption that the masses everywhere are no more than human resources needed for maintaining, increasing and modernizing the facilities that must assure the total and permanent enjoyment of the members of the privileged classes. The exploitation of natural and human resources has been the way for maintaining the efficient functionality of productive forces in order to determine the course of commercialized civilisation, owned, defined and modernised by some decision-makers inside the relevant ruling class. In 2008, many clever humans still try courageously to invent - through efficiently rationalized procedures - (empiric, scientific or magic) recipes for making a civilisation sustainable that definitely cannot be sustained any longer.

In 1971, Bertalanffy argued that "...what is badly needed is a timely image of man. Since the previous proud image derived from religion and philosophy does not serve modern needs efficiently, a new image should be synthesized ... I would contend that this is a very important business indeed – to find out what actually is human" (Bertalanffy). It is an essential concern that obliges researchers to follow the steps given by Lewis Mumford. "In terms of the currently accepted picture of the relation of man to technology, our age is passing from the primeval state of man, marked by his invention of tools and weapons for the purpose of achieving mastery over the forces of nature, to a radically different condition, in which he will have not only conquered nature, but detached himself as far as possible from the organic habitat".

With these new 'megatechnics', the dominant minority will create a uniform, all-enveloping, super-planetary structure, designed for automatic operation. Instead of functioning actively as an autonomous personality, man will become a passive, purposeless, machine-conditioned animal whose proper functions, as technicians now interpret man's role, will either be fed into the machine or strictly limited and controlled for the benefit of de-personalised, collective organisers of the machine.

Cosmic order was the basis of the new human order: The exactitude in measurement, the abstract mechanical system, the compulsive regularity of this 'megamachine', as I shall call it, sprang directly from astronomical observations and scientific calculations. This inflexible, predictable order, incorporated later in the calendar, was transferred to the regimentation of human components. Unlike earlier forms of ritualised order, this mechanised order was external to man. By a combination of divine command and ruthless military coercion, a large population was made to endure grinding poverty and forced labor at mind-dulling repetitive tasks in order to ensure 'Life, Prosperity and Health' for the divine or semi-divine ruler and his entourage.

However, "...certain missing components necessary to widen the province of the machine, to augment its efficiency, and to make it ultimately acceptable to the workers as well as to the rulers and controllers, were actually supplied by the other-wordly, transcendental religions: in particular by Christianity, since the seventeenth century ... (in) the Benedictine monastery ... all that the machine had hitherto been able to do only by making extravagant claims to divine mandate backed by large-scale military and paramilitary organisations was now done on a small scale, by small companies of men, recruited on a voluntary basis, who accepted work - indeed the whole technological order - not as a slave's curse but as part of a free man's moral commitment. Each able-bodied member of the monastery had an equal duty to work; each received an equal share of the rewards of work, though the surplus was largely devoted to buildings and equipment. Such equality, such justice, had rarely before characterized any civilised community, though it is a commonplace among primitive or archaic cultures. Each member had an equal share of the goods and food: and received medical care and nursing, plus extra privileges, such as a meat diet in old age. Thus the monastery was an early model of the 'welfare state' " ... "Now it became a model for cooperative effort on the highest cultural plane. Through its regularity and *efficiency* the monastery laid the groundwork for both capitalist organisation and further mechanisation: even more significantly, it affixed a moral value to the whole process of work, quite apart from its eventual rewards (Mumford, 1967).

3. Millions of humans have been trying to survive in what can only be called an inefficient manner

Our immediate ancestors were hominids who started to evolve in different ways. Their physiological features generated unique psychological features. We cannot imagine every detail of the impressions they had in their mind when one individual after another recognised that inevitably they would one day have to stop living. Different thoughts about the meaning of being alive led inevitably to the invention of *spirituality*. At first, it was no more than a sensible way of organising the actions required for the functioning of the emerging societies; later, it was aimed at justifying the cultural evolvement that some enlightened members chose for communal life. Apparently, many of our ancestors were not really conscious of the societal causality and effects that motivated every group to search for a way of creating its particular cultural ambience in their relevant natural environment.

More recently, during the last few centuries, after it became impossible to maintain the social atmosphere of the Middle Ages, some other minds, supported by accumulated empirical knowledge, invented peculiar actions (experiments) that gradually led to the emergence of *Science* as a human concern. This was a set of new procedures aimed at an in-depth comprehension of the causal interactions underlying the emergence of the diverse aspects of surrounding reality. Scientific knowledge - attained through trial and error and systematic experimentation - soon became another way of justifying the so-called development of every existing society. This kind of knowledge was used for changing a society's dynamic features by means of diverse kinds of technological devices and processes, supported by various methodological aids, which had been conceived and structured so as to make their utilitarianism more functional.

Spirituality and Science must be recognised as the outcome of two essential human concerns, each one introduced by clever minds into existing societies for the purpose of justifying the actions expected from their members. Both appear to be aimed basically at seeking how to render the presence of humans in time and in space more functional. Consequently, it may be considered proper - by definition and also by necessity - that these two concerns should complement each other because both are recognised as indispensable factors in the gradual integration of every personality. However, these two strictly human interests, being under the influence of civilisation's dynamics have instead disrupted the ways for improving human life and deviated it towards unfair competition that has increasingly dehumanised people's minds.

As Parra-Luna assumes, human personalities are indispensable for the emergence of humanitarian societies that may lead human minds to identify their sensible and responsible role on Earth. At present, this is only a hypothetical possibility in a few minds. What is actually obvious to billions of people is the use of spiritual interpretations and scientific assumptions - both presented as knowledge for the masses - for justifying the imposition of diverse kinds of religious dogmatism, compulsive militarism, civilian fundamentalism, homogeneous commercialism ... knowledge employed de facto for impeding the improvement of the diverse ways humans might perform, because it aims at an *efficient* homogeneisation of the utilitarian productivity of conglomerates composed of the “common” people.

Based on findings regarding the procedures implemented for governing numerous communities during the last 6,000 years, it may be assumed that in previous millennia, something similar happened. It means that the whole civilising adventure started when some enlightened minds invented the possibility for human beings to settle down in a chosen place. In recent years, the examination of archaeological and historical evidence offers testimonies of the creativity of the opportunistic utilitarianism of some humans who had found out how to transform natural circumstances to create "magnificent" human facilities, which were built by masses of people obliged to survive as submissive dehumanised entities. The dynamic features of the whole civilising adventure have been essentially determined by the performance of various leaders, each one supported by a small group of persons who managed to acquire privilege and to justify their dominance over the majority of people, who were expected to perform obediently and to remain passive.

Though most leaders intelligently structured their approach to accomplishing their personal intentions, today it appears obvious that they paid no attention to the societal consequences of their "brilliant" campaigns. These leaders - each supported by a minority group - managed to acquire privilege on their own, without noticing that they became inevitably engaged in the maintenance of increasingly contradictory civilising dynamics. Every oligarchy - without there being any need to measure their functionality attained through *efficient* corruption – engaged in actions for increasing its economic and social power in order to subjugate the masses located in the surroundings where each one organised its *effective operation*.

The successful manipulation of the masses everywhere around the planet, supported by various ideologies that refused to recognise human diversity, has caused homo sapiens sapiens to become an endangered species. It means that it will soon disappear from the planet, due to actions conceived by intelligently structured thinking that emerged in the course of millennia of "civilising" experiences; organised so as to ensure the effectiveness of the fallacious "civilisation" imposed on most societies; managed by successful minds engaged in mercantile and mercenary affairs. It will happen,

a) if humankind unconsciously maintains civilising trends, and regardless of the explosive growth of humankind, because some brains have been clever enough to invent more and more weapons for the annihilation of masses of people who have been considered during millennia as second class humans;

b) in spite of these masses having been considered useful at first because they could be employed as the slaves, servants and soldiers needed to permit the members of the privileged classes to enjoy living for the exclusive purpose of amusing themselves. Recently, over the last two centuries, members of the lower classes recognised as such by the members of the relevant oligarchy have become the "working class" needed for maintaining and increasing industrial production that was invented for making commercial business more and more profitable, certainly for the owners of enterprises;

c) because efficient and successful business has become the source of an increasing number of difficulties that are disrupting living circumstances on the planet Earth; the unexpected warming of the climate everywhere; an increasing pollution of natural resources; faster degradation of the quality of soil, water and air...;

d) because nowadays, the human world created by the development of science, technology and methodologies includes 1.300 million individuals who are starving; 500 million individuals not expected to live longer than 40 years; 1,000 million unemployed; 845 million of illiterates; 1,400 million without access to drinkable water; 30,000 children dying every day, 500,000 children and adults losing their eyesight due to lack of vitamins every year. Meanwhile, the World Bank informs us that 20% of the world population controls 80% of the gobal economy. In 1967, the income of rich countries was 37 times larger than that of the poor countries; today it is 74 times larger;

e) because neither the scientific, technological, methodological, sociological ... communities nor most governments seem to recognise the serious ethical, ethological and ecological crisis into which the Earth has been plunged as a result of civilisation. The ambitious axiological approach of Parra-Luna (above) for measuring an overall concept of Efficiency is expected to fail precisely because of the opposition of these forces.

However, a *different perspective* might be created if groups of dissidents everywhere succeeded in learning to *think differently*. The first step would have to be a recognition of the nearly catastrophic situation that has been systematically built on the planet through the gradual aggregation of the effects of actions based on intelligently perverse reasoning - narrow-minded, prejudiced reasoning unable - both theoretically and practically - to create awareness of the crisis and to believe that problems may be overcome is clamouring for sustainable development.

The global panorama at present appears to be on an extinction curve after wild nature has been transformed – in the course of relatively few millennia – into increasingly disrupted environments chaotically interrelated through mercantile and mercenary channels, where the masses of homo sapiens sapiens still manage, though unconsciously, to survive. The members of only one species, the human one, are now consuming, wasting, or diverting approximately 45% of total net biological productivity on land and using more than 50% of the renewable fresh water. A large proportion of living species indispensable for the survival of the human race have been driven to extinction due to the careless and irresponsible means humans use to accomplish their ends.

The immediate perspective for humankind – created by highly civilised specimens – is even more dangerous, due to the threat of nuclear war, which is encouraged - quite contradictorily - by scientific and technological minds and technical brains who have been increasing the number of weapons. The more than 50,000 nuclear warheads in existence today could be used efficiently for destroying the Earth's population 60 times over. We can never be sure that we will be able to manage the disaster.

If the global costs and scientific resources used for the defence of the most highly civilised countries (published from time to time by way of efficient strategies developed for commercial marketing) were used for maintaining ecological conditions, a sustainable civilisation might be built and maintained. Financial resources that could be used for providing sanitation and clean water to all the deprived peoples, in order to prevent the spreading of diseases, are used instead for increasing the potential of warfare against invented enemies and also for erasing the increasing number of dissidents. Health systems all over the world are not really concerned with the management of disease through health promotion, which would increase the nutritional, environmental, and preventive facilities needed for attending properly to five billion people. Instead, the pharmaceutical industry - which play a decisive role in all health systems thanks to various corrupt methods and technical procèdures – focus on the efficient marketing of medicines that most people – especially the poor – cannot afford. In the U.S.A., the pharmaceutical industry does not seem to be concerned with anything else except making financial profits from selling drugs anywhere and everywhere.

"Two crucial developments in U.S. constitutional jurisprudence—the grant of Bill of Rights protections to corporations, and the extension of First Amendment protections to commercial speech—have enabled corporations to invoke the First Amendment to defend their right to have goods, so long as they are legal, by almost any means short of outright lying or clear deception. ...Drug companies devote much more money, and time, to influencing those with the power to prescribe medicines—for instance, billions of dollars in the United States, which is purportedly several times what is spent on direct-to-consumer marketing. The most important element of the marketing onslaught directed at doctors is "detailing" — the activities of the sales representatives who visit doctors constantly, and provide free lunches, free pens, free charts and other free goodies (including, very importantly, free samples). The average primary care physician sees drug detailers more than five times a day. When a sales rep walks into a doctors office, he or she knows a lot about that doctor—including exactly what medicines the doctor prescribes, and in what quantities. How can this be?

Pharmaceutical companies purchase the information from data-mining companies, the largest of which is IMS Health. Pharmacies track what drug is sold to each customer. IMS buys the data from the pharmacies, deletes all patient names, combines it with data that enables the identification of prescribers for each prescription, and aggregates the information. ... And, as the New York Times explained, quoting an e-mail message from a pharmaceutical executive to company salespeople, they use the data to "hold [doctors] accountable for all the time, samples, lunches, dinners, programs and past preceptorships that you have paid for and get the business!" The sales reps obviously do not have punitive power over the doctors, but they use the prescribing information to exploit and manipulate the social ties built on the giving relationship." (Weisman, 2008)

The modernisation of societies has been recently re-structured as a trajectory for emphasising commercial returns, economic growth, increasing wealth, societal success... irrespective of the unexpected consequences modernisation produces by subordinating the majority of people and distorting their thinking when they try to identify their needs. Only a global consideration of the whole system of values, to be performed for the benefit of the mankind, would make sense, but we cannot be optimistic about its application because economic, religious and military powers are not ready to accept the challenge.

4. An efficient approach to a responsible humanisation of dehumanised humanity

The nearly catastrophic situation that humankind is facing at present is an urgent problem clearly identified by enlightened minds over the last century. Taking into account some ideas of Bertalanffy, two world wars showed the people of this epoch, whose hope it was to create a happy future for mankind by means of mechanics and technology, that their hope had been misplaced. We have to find a new orientation in the world, to learn again to believe in its rationality. Our perception of the world has become chaotic; we have to reconstruct it in us as a cosmos. The technical age is about to become disgusted with itself - let us hope that it will be followed by an organismic one able to open new doors for the future of mankind. (Pouvreau & Drack, 2007).

"Without jumping to conspiracy theories, or citing the illegal activities which now constitute the world's biggest industry, we can at least say that humankind now manages its own affairs with breathtaking incompetence...": "...Small tribes managed themselves very well indeed, and without destroying their habitats..." (Bear, 1993). As Raven (1995) stated for the last century, "the problems we face stem from the pervasive and perverse persistence of the 'American Dream'. Not only has this proved to be a mirage even in America itself, but long before all the world's people – or even just the people of China – could achieve it, the planet would be laid to waste. Apart from anything else, living like Americans depends on endless resources sucked in from other parts of the globe. The west consumes more than three quarters of all the world's metals and energy and causes the bulk of the pollution of the soils, seas, and atmosphere. It accounts for two thirds of all the greenhouse gases and three quarters of the sulphur and nitrogen oxides that cause acid rain".

In spite of numerous warnings about the increasingly dangerous situations generated by civilising dynamics, the crisis has not been yet sufficiently recognised in 2008; instead, more and more civilising absurdities continue to be accomplished through the decision-making maintained by the privileged societies supported by the ingenuous passivity of submissive masses. The crisis is clearly characterised by numerous problematic situations:

1) Economic perspectives disturbed by unemployment, poverty, misery, hunger, starvation, crime ...

2) Social situations infected with racism, xenophobia, non-declared apartheid, religious and ethnical intolerance ... weakening of human motivation and poor social solidarity ...

3) Economic possibilities disrupted by the waste of valuable material resources, the depletion of non-renewable mineral and energy reserves, wastage and underemployment of human capacities; mismanagement of human organisations ...

4) Physiological and mental disruptions in an increasing number of humans, emergence of new diseases and aggravation of old ones (aids, Alzheimer's disease, skin cancer ...), an increased incidence of drug-addiction and neuroses; attacks on man's genetic heritage ...

5) A growing traffic in arms and weapons used for destroying cultural accomplishments; export of inflation and recession; export of capital from poor countries to highly industrialized ones; exportation of rubbish and technological garbage from rich countries to poor ones; unreasonable, unjust and dishonest affairs ...

6) Environmental circumstances exacerbated by damage to vital properties of the natural elements, deterioration of the quality of water, air, soils; ongoing desertification caused by the increasing exploitation of forests, jungles, lands, oceans and natural beauties; acid rain ... deterioration and destruction of the habitats of many living species, a growing number of endangered species ...

7) Changes of climate caused by greenhouse effects due to pollution of air and deterioration of the ozone layer (ecosphere) ...

8) Growing corruption, hypocrisy, cynicism ... and even frivolity

However, to deal consistently with these situations, we must identify the causes of the crisis, which appear to be due to creativity encouraged by:

1) The pervasive and perverse persistence in trying to validate civilisation trends.

2) The belief in endless resources waiting to be sucked in from the globe.

3) Unilateral aggregation of regional, national and international organisations for assuring the success of global marketing.

4) The use of money for producing and selling weapons for annihilating the rebellious populace who dares to resist and fight against empowered rulers who are using democratic procedures for maintaining the division between the wealthy minorities and the miserable masses and to assign profitable contracts to corrupt enterprises.

5) The cynical justification of the First World War of the XXI century, which used violent terrorism to guarantee ownership and exhaustive exploitation of

oilfields and lands for producing the bio-combustibles needed for maintaining and increasing unsustainable industrialism and facilities for private transportation, as well as allowing big corporations to create systems for rendering Global Free Marketing functional, with the aim of increasing - ad infinitum - their profitable involvement in business, financing and economic transactions. This is a generalised war cynically supported by corrupted economic, social and cultural procedures aiming to destroy borders and eliminate nationalism, in order to be able to use governments for the systematic and efficient homogenisation of more and more minds.

Despite warnings, the perspective that generates warfare has been successfully integrated. "It is commonplace to say that the utopia of progress which has guided Western science and technology from its beginning up till now has faltered in the modern world, when control of physical forces has led to the menace of atomic annihilation, and society has become meaningless and unhappy in the midst of plenty".

In a letter written to Henry Osborn Taylor in 1905, Henry Adams remarked "At the present rate of progression since 1600, it will not need another century or half century to tip thought upside down. Law, in that case, would disappear as theory or a priori principle and give place to force. Morality would become police. Explosives would reach cosmic violence. Disintegration would overcome integration» Henry Adams did not live to observe fascism: he anticipated it. He knew that the detonators of violence and destruction were present in every part of the social structure of Western society (Munford, 1944).

So far, it cannot be denied that empowered humans have chosen to learn and develop Machiavelli's ideals in order *to govern by means of force, terrorism and corruption.*

To overcome the actual crisis, we must use the potential of human intelligence through holistically responsible concerns, bearing in mind that the successful (?) civilised trajectory over the last few centuries has been determined by the intelligent reasoning practised by numerous scientists who were used to perform analytically and axiologically:

"wholeness" and its derivative, "organisation", appear as the categories of life ... since the end of the 19th century, (being) naturally related to the rise of "*Lebensphilosophie*" and anti-"mechanicism" ... from the 1920s on, *Ganzheit* ("wholeness") becomes an avatar of ... ("the one and the whole") ... protesting against the fragmentation of knowledge and man ... . Bertalanffy considers himself as an heir of (the) alleged "tradition", roughly unified by a vision of the universe as a hierarchically and dynamically organised whole in which analogy works as the unifying principle...

> In the 1920s and 1930s, *Ganzheit* ("wholeness") works as a metaphor organising research, widening its meaning more and more and connecting different fields... the *Ganzheit* thinkers rejected a "mechanistic" science and a "mechanised" life, an "atomistic" epistemology and the atomisation of individuals.

> Bertalanffy ... (tried)... to conciliate the recognition of the fundamental significance of *Ganzheit* with science: The confidence crisis of science must not lead to a Mephistopheles-like contempt of reason and science (Bertalanffy 1930). Organic *Ganzheit* is neither a metaphysical concept nor an asylum of ignorance, but a problem that can and must be investigated using the methods of exact science (Bertalanffy 1930 and 1937a). The organismic conception, first developed from biology, could eventually provide a general world view (Bertalanffy 1934b). (Pouvreau and Drack, 2007)

It is necessary to learn to deal with structured wholes on an inter-disciplinary basis while recognising that the various variables may cause the emergence of casual effects. This requirement could be implemented today through the possibilities offered by Systems Thinking, Cybernetic Interactions and Information and Communication Technology, although it would be necessary to *study the available research results* on how to use these methodological tools for comprehending the crisis holistically, and afterwards for learning how to overcome it. Meanwhile, it is indispensable to comprehend why and how these tools have been used to support the fallacious assumption that civilisation can be sustained forever by efficient actions for business' sake through sustainable development.

This *Methodological Research* essentially comprises seven concerns (described briefly below):

1) A holistic approach to the crisis, in order to facilitate the identification of its causes, which is the indispensable antecedent for creating the atmosphere needed for learning: (a) the harmonious maintenance of the features of the Ecosphere; (b) a consistent development of the Biosphere; (c) the creation and maintenance of the humane functionality of the Homosphere. It is quite difficult work, because to deal holistically with something means to take into account all the factors involved, which makes it necessary to organise this task so as to allow the minds involved to examine their different appraisals continuously, in a system of collaboration and mutual support needed for reaching an agreement.

2) The recognition of every human concern as an open system, which means to take note well in advance of how the diverse aspects of the surroundings will or may affect the functionality of the system and also how the surroundings will or may be affected inconveniently by the system. Considering humankind as an open system will help us learn how to prevent inconveniences that might happen in the ecosphere, in the biosphere and in the homosphere. It will also help to define the requirements of human performance in the light of ecological, biological, ethological and ethical criteria.

3) Proper identification of the various kinds of systems that humans develop in order to distinguish particular ways of dealing with societal, economic, political, cultural, ideological... concerns.

4) Examination of ways in which humanitarian purposes are integrated into the design of a system, bearing in mind that many systems have been conceived and structured for reaching arrogant, perverse, inhuman purposes supported by monetary, dogmatic, opportunist interpretations of particular

aspects of the human world. A kind of axiological empirical approach, such as the one proposed by Parra-Luna (above,) would be necessary in order to see the abandonment of the value "Nature Conservation".

5) Examination of when and how systems should be created that are aimed at the manifestation of unity through diversity.

6) Development of educational systems that allow human beings to develop their personalities with dignity, while remaining committed to ethical, ethological, biological and ecological responsibility.

7) In-depth examination of how to design the self-organising possibilities of systems needed by a particular community, whose functionality should be basically structured by means of (a) negative feed-back interactions aimed at correcting any possible deviation and (b) positive feed-forward interactions for encouraging the development of humane responsibility.

The above paragraphs should probably be used as the conclusion of my speculations on recognising the various and serious disruptions that narrow-minded interpretations of this concept have caused as a result of industrialism aiming to consolidate its impossible, chaotically structured civilising efforts. I shall therefore round off this brief reflection by evoking some of the deep and carefully expressed thoughts of *Mahatma Gandhi (1869 –1948).*

4.1 "...Modern civilisation? That would be a good idea"

Gandhi recognised that the anthropocentric view of the world adopted by the developed countries has been the source of its perilous trajectory. In the anthropocentric view, humans are (a) the center of the universe, they are (b) the crown of creation, therefore they have (c) absolute rights over Nature, entitling them to take whatever they want freely from their natural surroundings.

Anthropocentrism recognises de facto that (a) desires and wants of humans have no inherent limits, that (b) one desire leads to another; these two assumptions explain the obsession with the satisfaction of desires, which leads inevitably to consumerism.

This view allows humans to believe that Nature is an inorganic mass of material that can be reduced to environment or resources needed for the efficient manifestation of industrialism; the industrialism that would permit humans to strip the world bare like locusts.

Gandhi claimed that the anthropocentric view of the world should be replaced by the cosmocentric view, which means that humans are located in the wider cosmos where they should relate to its other members as equal tenants. In this appraisal, Nature is a self-regulating system that humans should respect and fit into.

Gandhi's *truth force = satyagraha*, first developed in South Africa against the evils of apartheid, was employed later to empower the people of India to wage their struggle for freedom from colonialism. But thinking about the need of getting rid of a centralised, exploitative and violent system of governance led Gandhi to conceive the principles of a nonviolent social order for a truly free India. The Gandhian Trinity comprises three parts: *Sarvodaya, Swaraj, Swadeshi:*

1) *Sarvodaya = Upliftment of All* means the care of the Earth, of animals, forests, rivers and lands, as well as people, as opposed to democracy as a tool for maintaining the division between the rich minority and the less well-off majority of people ...

2) *Swaraj = Self Government* means social transformation through a small-scale, decentralised and participatory structure of government, and self-transformation, self-discipline and self-restraint on a personal level ...

3) *Swadeshi = Local Economy* means production by the masses, as opposed to mass production. Work is as much a spiritual necessity as an economic one. Every member of society should be engaged in manual work. Manufacturing in small workshops and adherence to arts and crafts feed the body as well as the soul ...

Following the Gandhian way of thinking shows how Western democracy favours the few who have capital and helps them to maintain the majority - who should work, buy and consume without complaint - in submission, and why sustainable development is a wistful fantasy with a peaceful and a violent side. Development comes with the hubris of globalisation; flies its banner of progress, convenience and scientific technological advancement that ultimately spells obsolescence; renders obsolescent all traditional peoples, their cultures, the nature of their places, their sense of community. Sustainable ideas are ideas for non-violence and peace; decentralisation and diversity for keeping alive the physical and cultural environment of people, their places, their waters, their traditions of birthing and dying, of marrying or parting, of eating and defecating.

Conclusion

Gandhi practiced what he preached, being convinced that it is necessary for every human to live the change that he wants to see in the world. Therefore the right conclusion to this cogitation, which was not conceived as if it were simply guesswork, comprises two actions that must be examined and then be implemented by scientists responsibly engaged in humanising the dehumanised homosphere.

Firstly, "two recent reports of the Food Climate Research Network and the United Nations Food and Agricultural Organisation show that the agricultural livestock sector emerges as one of the top three most significant contributors to the most serious environmental problems. ...(it) is responsible for "18% of greenhouse gas emissions measured in carbon dioxide equivalent, which is a higher percentage than that attributable to transport" ...raising animals for food generates more green house gases than all the cars and trucks in the world combined. The livestock sector accounts for 37% of methane emissions too, from the enteric fermentation of ruminants. ...twenty three times more potent at trapping heat in our atmosphere than carbon dioxide. This is to say nothing of rainforest depletion for ranching, fossil-fuel use for pesticides, fertilizers and farm equipment, water usage and pollution from livestock excreta ("livestock in the USA produce 130 times more excrement than the entire human population every day") or indeed the abhorrent cruelty endemic in the massive-scale factory farms that most developed countries deem economic. If everyone who eats meat made a conscious decision to reduce their intake by 50% this would have a huge, immediate, positive impact in reducing carbon emissions" (Lorna Howarth (16)). 90% would be better.

Secondly, "Gandhi transformed defecation and urination into matters of profound political and economic importance, deserving the deepest philosophical and spiritual investigation. He elevated these fleshly matters into critical questions about democracy, autonomy and liberation from unjust social practices and institutions. ... Today's elites – the developed peoples of the global economy – share the addictive privilege and convenience of having their 'wastes' transported out of sight and out of mind – first made possible by Mr. Crapper (designer of the first flush toilet for the King of England). ... Are we taking care of our shit by flushing it away? ... pause (is) needed for truly addressing the questions of how we use, abuse and waste water, rendering it toxic as well as globally scarce. As the privileged elites flush away their daily wastes, the social majorities stand for endless hours in long lines and queues to get their daily bucketful of government-allocated treated toxic waters. Water wars are already competing with oil wars. ...the media are not hiding these water wars... (but are) failing still to connect the dots, ... they have not yet traced these global water wars to their source in personal, private flush toilets: failing to link the death of the oceans to our addiction to the 'progress' that demands that our excrement must be removed out of sight, out of mind. More than 40% of the water available for domestic purposes is used everywhere for the transportation of shit. And this causes problems of health and serious pollution of soil and water. Mixing three marvellous substances (water, urine and shit) we concoct a poisonous industrial cocktail. ... the 'waste Gods of Development' make everything in Nature – glaciers, ice caps, rivers, even oceans and certainly aquifers and ground waters – *toxic, scarce or non–renewable*. ...it makes logical sense to stop the daily ritualised production of black waters; to wean ourselves from our addiction to transporting our shit and urine away in thousands of miles of sewage pipes to treatment plants."

"My own awakening to Gandhi's teachings on human shit and urine started not in my native India, but instead in my adopted village in Mexico. There, a 'commoner' is pioneering a movement for the Ecological Dry (urine-separating) Toilet. Learning simple, inexpensive and commutable designs and principles of 'people's science' from Guatemalan and Vietnamese peasants, César Añorveis is protecting water, humanure and urine... A bucketful of soil collected from our own backyards, combined with some lime is all we need to disconnect ourselves from unsustainable practices." Madhu Suri Prakash (2008).

References

Beer S. (1993) "World in Torment: A Time Whose Idea Must Come". *Kybernetes* 22 (6), 15 – 43.

Bertalanffy, L. (1967) *Robots, Men, and Minds: Psychology in the Modern World.* New York: George Braziller.

Fowler, H.W. (1926) *A Dictionary of Modern English Usage*. UK: Oxford University Press.

Funk & Wagnalls (1970) *Standard Dictionary of the English Language.* International Edition Volume One & Two.

Huxley, A. (1980) *Culture and the Individual.* In *Moksha*. ed.by. Horowitz M. and Palmer C., Chatto & Windus.

Kuper A. & Kuper J. (1985) *Social Science Encyclopedia.* Routledge & Kegan Paul.

Williams, A & Anderson R, (1975) *Efficiency in the Social Services.* Oxford: Basil Blackwell.

Longman (1978) *Dictionary of contemporary English.* London.

Madhu Suri Prakash (2008) "Honouring Gandhi: a reflection on defecation and development". *Resurgence* No. 246, January-February.

Merriam-Webster, Thesaurus (1978). Pocket Books.

Moliner M. (1984) *Diccionario del Uso del Español.*

Mumford, L. (1944) *The Condition of Man.* New York: Harcourt Brace Jovanovich.

Mumford, L. (1967) *The Myth of the Machine.* London: Secker & Warbur. Oxford English Readers Dictionary (1952)

Pouvreau D, Drack M. (2007) "On the history of Ludwig von Bertalanffy's 'General Systemology', and on its relationship to Cybernetics – Part I:Elements on the origins and genesis of Ludwig von Bertalanffy's 'General Systemology'". *International Journal of General Systems* 36, 281-337.

Raven, J. (1995) *The New Wealth of Nations: A New Enquiry into the Nature and Origins of the Wealth of Nations.* New York: Royal Fireworks Press; Sudbury, Suffolk: Bloomfield Books.

Scott J.A. (1950) *The Measurement of Industrial Efficiency. London : Sir Isaac Pitman & Sons.*

Weissman, R. (2008) *The First Amendment Gone Wild: Big Pharma's 'Right' to Find Out What Doctors Are Prescribing.* Available from <http://www.healthyskepticism.org/global/library/item/12370>

Williams, A & Anderson R. (1975) *Efficiency in the Social Services.* Oxford and London: Basil Blackwell and Martin Robertson.

8

Concluding remarks

by **Francisco Parra-Luna** and **Eva Kasparova**

All the authors participating in this volume are conscious of the central role that the human being plays in all kinds of organisations, even though some of them deal with very specific technical issues apparently little related to this humanistic perspective.

But it is a known fact that "efficiency theories", "innovation evaluation", "environmental damage caused by air transport", "non-profit organisations" and, naturally, "radical criticism which analyses the world social situation", only strive at the end to satisfy the interests, needs, desires and motivations of the human beings that belong or are related to these organisations.

That is why it seems absolutely necessary to start with an inventory of basic and derived human needs applicable to all societies and organisations in time and space. Maslow's well known scheme of needs served as a first step. Moreover, the empirical possibility of dealing with these universal human motivations exists and has generally been accepted and enshrined in the U.N. Universal Declaration of Human Rights in 1949, when most of nations agreed that it was necessary for all the countries of the world to achieve adequate levels of physical and mental health, economic development, to improve and generalize education, and, at the same time, to reach adequate levels of distribution of goods, political and social freedoms, as well as peace, order and security among others. In addition, today, after many international conferences on environmental problems, the protection of nature is seen also as an important necessity for the human being.

In other words, we cannot overlook peoples' motivations at work, leisure in their free time, rest when they are tired, or to be involved in social or political participation. Because through any of these activities, what they are looking for is to obtain adequate levels of Health, Wealth, Security, Nature Conservation, Justice, Freedom, and so on. Even if we repeat the reasoning of previous chapters, it must be stressed that we are faced with the inescapable need to achieve these values, because we, as humans, are precisely made up of these deficiencies. We, as biosocial individuals, could not pursue other goals even if we wanted to, because it seems that we are biologically and socially programmed or predetermined to try to be happy, and happiness is nothing other than living according to the system of values each individual thinks is the best and most appropriate at each moment of his/her life. In fact, in our everyday existence we do nothing but focus on some values more than on others, depending on the circumstances. That is why some people prefer, e.g., to buy an expensive car (thus trying to increase the value of "Social Prestige", even at the cost of having less money (a lower level of "Wealth"), or prefer to spend most of their time reading books because they think that increasing culture (value of "Knowledge") helps gain independence (or Freedom), or prefer to work in an company that helps its employees achieve maximum personal development (value "Quality of Activities") even if it is a riskier company to work (lower level of the value of "Security"). That is why, in this sense, each one of us is nothing but a kind of "dynamic axiological profile".If that is so, it seems obvious that any effort to define, measure, and quantify the levels of achievements of these values, starting by using a comparable pattern of reference for them, is something that could help humanity tackle the many grave problems we encounter. Maybe a good way of dealing with any important problem in social life is to identify its axiological structure and the way in which we can operationalise and quantify it. Could we then talk about Organisational Efficiency without talking about the "systems of values" that any organisation perform for the benefits of the people which depend on it? It is therefore surprising to see still many scientific works dealing with the subject but where the analysis of values is absent. The humble purpose of this book is only to open one way to this target.

Index of names

Francisco Parra-Luna is Full Professor (Emeritus) at the University of Madrid, Spain and has been the Director of the University Institute for Human Resources.

Campus de Somosaguas, 28223, Madrid, Spain

Tel.: 0034 6389731

Fax: 0034 6389731

Móvil 670 649 637

E-mail: parraluna3495@yahoo.es

PhDr. Eva Kasparova, Ph.D. is a senior lecturer at the Faculty of Business Administration, University of Economics in Prague.

Department of Managerial Psychology and Sociology

W. Churchilla 4

130 67 Prague 3 – *Žižkov*

Czech Republic

Tel.: +420 224098330

GSM: +420 737088608

Fax: +420 224098633

E-mail: kasparov@vse.cz

Antonio Sanchez holds a Master in Aerospace Engineering (1993) and a PhD in Engineering Physics (1995) in the Department of Applied Mechanics and Engineering Sciences.

Universidad Complutense de Madrid.

Facultad de Ciencias Políticas y Sociología

Departamento de Sociología

Campus de Somosaguas, Madrid

Contact telephone 34 650 867833

E-mail: asanchezsu@terra.es

Chaime Marcuello-Servós is a member of GESES (Third Sector's Social and Economic Studies Group) and Professor at the Psychology and Sociology department at Zaragoza University (Spain), (http://geses.unizar.es/)
EUES-Violante de Hungría, 23, 50009 Zaragoza, Spain
Phone: 34 976 76100
Fax: 34 976 761029
E-mail: chaime@unizar.es

Elohim Jiménez López is a senior researcher at the Bertalanffy Center for the Study of System Science(s) in Vienna.
Missongasse 9E, 3512 Mautern a. d. Donau, Austria
Phone: 43 2732 20228
Fax: 43 2732 20228
E-mail: elohimjl@gmail.com

Prof. Emer. Dr. Dr. Matjaz Mulej, 1941, is retired from University of Maribor as Professor of Systems and Innovation Theory. He published +1.700 contributions in close to 50 countries and was visiting professor abroad for 15 semesters. He is a member of three international academies of science, including IASCYS (vice-president, in 2010-2012 president). Now he chairs the scientific research of IRDO Institute for development of social responsibility in Maribor; on behalf of it he co-edited ten books and issues of systems-science journals, published in 2013 and 1014 with about 500 authors from +30 countries. 'ResearchGate' reports that several thousand persons have demonstrated interest in his work.
Razlagova 14, Maribor,2000, Slovenia
Phone: 386 2 2290262
Fax: 386 2 2516681
E-mail: mulej@uni-mb.si

Peter Fatur, MBA, M.A., used to work in several companies in Slovenia and has dedicated his research to innovation processes in 'Submit Your Idea' style.

Two years ago he joined faculty of University of Primorska, Faculty of Management, Koper.

Dr. Jožica Knez-Riedl is Assoc. Prof. at the University of Maribor, Faculty of Economics and Business, lecturing and researching the fields of business economics, environmental economics, managerial economics, economics of projects, creditworthiness of a firm, and corporate social responsibility. Pedagogically she is engaged at both undergraduate and postgraduate levels, including MBA and specialists studies (also at Economic faculty of University of Rijeka, Croatia). She is author and co-author of scientific and professional monographs, published in Slovene, English and German language, of several cited articles in international scientific reviews, and of papers at international conferences (over 60 papers).

Andrej Kokol, M.B.A., is employed in Unior Zreče, Slovenia, since 1996 with several engeeniring related duties, since 2004 head of 'Submit Your Idea? Team, since 2001 member of the governing board, co-established of two small enterprises, and Ph.D. student.

Damijan Prosenak, MBA, born in 1967, is a sole entrepreneur. He obtained the Bachelor's Degree on Faculty of Electrical Engineering and Computer Science, Maribor in 1992. Afterward he decided to manage the existent family business. In 2005 he obtained Master's Degree in Business Administration on Faculty of Business and Economics, Maribor. His main fields of research interest are innovation management, marketing, networking and social responsibility and their requisitely holistic application in business and other practice.

Dr. Branko Škafar, is general manager of Saubermacher-Komunala Murska Sobota. He received a prestigious award from the Chamber of Commerce and Industry of Slovenia for exemplary achievements of major significance in the field of business for 2002. He is a lecturer at the School of Economics in the technical college at Murska Sobota and a part-time guest lecturer in MBA studies at the Faculty of Economics and Business, University of Maribor. He has published several articles at home and abroad, as well as 4 books. His basic areas of practice and research are quality, business excellence, innovation, environmental protection, organisation and management.

Dr. Zdenka Ženko is Assistant Professor of Innovation Theory and System Theory at Faculty of Business and Economics, University of Maribor. Her research interest includes dialectical system theory applied in economic fields. Her research includes systemic approach to innovation management; mostly for small and medium enterprises prevalent in Slovenia and comparative studies of most developed innovative management models.

www.ingramcontent.com/pod-product-compliance
Lightning Source LLC
Chambersburg PA
CBHW052141070326
40690CB00047B/1346
* 9 7 8 1 8 4 8 9 0 1 2 6 1 *